Teacher's guide to Book 7C

CAMBRIDGE
UNIVERSITY PRESS

CAMBRIDGE UNIVERSITY PRESS

Cambridge, New York, Melbourne, Madrid, Cape Town, Singapore, São Paulo

Cambridge University Press
The Edinburgh Building, Cambridge CB2 2RU, UK

www.cambridge.org
Information on this title: www.cambridge.org/9780521537919

© The School Mathematics Project 2003

First published 2003
4th printing 2006

Printed in the United Kingdom at the University Press, Cambridge

A catalogue record for this publication is available from the British Library

ISBN-13 978-0-521-53791-9 paperback
ISBN-10 0-521-53791-6 paperback

Typesetting and technical illustrations by The School Mathematics Project and Jeff Edwards
Photograph on page 207 by Graham Portlock
Cover image Getty Images/Randy Allbritton
Cover design by Angela Ashton

Contents

Introduction *5*

Mental methods and starters *7*

1 First bites *12*

2 Symmetry *27*

3 Number skills *38*

4 Growing patterns *44*

5 Test it! *50*

6 Number patterns *53*

7 Angles and triangles *58*

8 Fractions *64*

Review 1 *68*

Mixed questions 1 *68*

9 Action and result puzzles *69*

10 Chocolate *71*

11 Health club *73*

12 Balancing *80*

13 Multiples and factors *82*

14 Work to rule *86*

15 Decimals 1 *90*

16 Gravestones *97*

Review 2 *102*

Mixed questions 2 *103*

17 Area and perimeter *104*

18 Negative numbers 1 *108*

19 Spot the rule *111*

20 Chance *115*

21 Translation *121*

22 Stretchers *124*

23 Number grids *126*

24 Constructions *136*

25 Comparisons 1 *138*

26 Further areas *144*

Review 3 *148*

Mixed questions 3 *148*

27 Inputs and outputs *149*

28 Decimals 2 *156*

29 Investigations *162*

30 Parallel lines *171*

31 Percentage *175*

32 Think of a number *180*

33 Quadrilaterals *186*

34 Negative numbers 2 *192*

35 Comparisons 2 *195*

Review 4 *200*

Mixed questions 4 *200*

36 Know your calculator *201*

37 Three dimensions *205*

38 Finding formulas *212*

39 Ratio *219*

40 Using a spreadsheet *224*

41 Functions and graphs *226*

Review 5 *229*

Mixed questions 5 *229*

The following people contributed to the writing of the SMP Interact key stage 3 materials.

Ben Alldred	Ian Edney	John Ling	Susan Shilton
Juliette Baldwin	Steve Feller	Carole Martin	Caroline Starkey
Simon Baxter	Rose Flower	Peter Moody	Liz Stewart
Gill Beeney	John Gardiner	Lorna Mulhern	Pam Turner
Roger Beeney	Bob Hartman	Mary Pardoe	Biff Vernon
Roger Bentote	Spencer Instone	Peter Ransom	Jo Waddingham
Sue Briggs	Liz Jackson	Paul Scruton	Nigel Webb
David Cassell	Pamela Leon	Richard Sharpe	Heather West

Others, too numerous to mention individually, gave valuable advice, particularly by commenting on and trialling draft materials.

Editorial team	**Project administrator**	**Design**	**Project support**
David Cassell	Ann White	Pamela Alford	Carol Cole
Spencer Instone		Melanie Bull	Pam Keetch
John Ling		Nicky Lake	Jane Seaton
Paul Scruton		Tiffany Passmore	Cathy Syred
Susan Shilton		Martin Smith	
Caroline Starkey			
Heather West			

Special thanks go to Colin Goldsmith.

Introduction

Teaching approaches

SMP Interact sets out to help teachers use a variety of teaching approaches in order to stimulate pupils and foster their understanding and enjoyment of mathematics.

A central place is given to discussion and other interactive work. In this respect and others the material supports the methodology of the *Framework for teaching mathematics*. Questions that promote effective discussion and activities well suited to group work occur throughout the material.

Some activities, mostly where a new idea or technique is introduced, are described only in the teacher's guide. (These are indicated in the pupils' book by a solid marginal strip – see below.)

Materials

There are three series in key stage 3: *Books 7T–9T* cover up to national curriculum level 5; *7S–9S* go up to level 6; *7C–9C* go up to level 7, though schools have successfully prepared pupils for level 8 with them, drawing lightly on extra topics from early in the *SMP Interact* GCSE course.

The year 7 books share much common material – a benefit where mixed attainment groups or broad setting are used for an initial settling-in period, or where the school covers topics in parallel to ease transfer between sets. To help you with your planning, links to common and related material – between *Book 7T* and *7S*, and between *7S* and *7C* – are shown to the right of unit headings (in both the pupils' book and the teacher's guide); for example, unit 2 of *Book 7S* has the links '7T/5, 7C/2', meaning there is common or related material in unit 5 of the less demanding *Book 7T* and in unit 2 of the more demanding *Book 7C*.

All three year 7 books start with a collection of activities called 'First bites', designed to help you get to know pupils and to give them an enjoyable and confident start in mathematics at secondary school.

Pupils' books

Each unit of work begins with a statement of learning objectives and most units end with questions for self-assessment.

 Teacher-led activities that are described in the teacher's guide are denoted by a solid marginal strip in both the pupils' book and the teacher's guide.

Some other activities that are expected to need teacher support are marked by a broken strip.

Where the writers have particular classroom organisation in mind (for example working in pairs or groups), this is stated in the pupils' book.

Resource sheets

Resource sheets, some essential and some optional, are linked to some activities in the books.

Practice booklets

For each book there is a practice booklet containing further questions unit by unit. These booklets are particularly suitable for homework.

Teacher's guides

For each unit, there is usually an overview, details of any essential or optional equipment, including resource sheets, and the practice booklet page references, followed by guidance that includes detailed descriptions of teacher-led activities, advice on difficult ideas and comments from teachers who trialled the material.

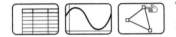

There is scope to use computers and graphic calculators throughout the material. These symbols mark specific opportunities to use a spreadsheet, graph plotter and dynamic geometry software respectively.

Answers to questions in the pupils' book and the practice booklet follow the guidance. For reasons of economy answers to resource sheets that pupils write on are not always given in the teacher's guide; they can of course be written on a spare copy of the sheet.

Assessment

Unit by unit assessment tests are available both as hard copy and as editable files on CD (details are at www.smpmaths.org.uk). The practice booklets are also suitable as an assessment resource.

Mental methods and starters

Mental methods

To be successful in developing and using mental methods of calculation the pupil must be able to recall rapidly a range of basic number facts. These include

- addition and subtraction facts to 20
- multiplication tables and the corresponding division facts
- using place value to multiply and divide by 10, 100, 0.1, and so on

Some strategies for mental calculation are described below.

Using an imagined number line

The number line is useful for visualising additions and subtractions.

Example $13.5 - 5.7$ (thought of as $5.7 + ? = 13.5$)

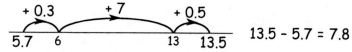

$13.5 - 5.7 = 7.8$

Using multiplication and division facts and place value

Example $0.4 \times 7 = (4 \times 7) \div 10 = 28 \div 10 = 2.8$

Re-ordering

A mental calculation is often made easier by re-ordering.

Examples $5.3 - 1.8 + 2.7 = 5.3 + 2.7 - 1.8 = 8 - 1.8 = 6.2$

$4 \times 17 \times 5 = 4 \times 5 \times 17 = 20 \times 17 = 340$

Using factors

This can sometimes make multiplication or division easier.

Examples $35 \times 12 = 35 \times 2 \times 6 = 70 \times 6 = 420$

Dividing by 18 is equivalent to dividing by 2, then by 9

Dividing by 5 is equivalent to dividing by 10, then doubling

Doubling and halving

In a multiplication it can help to double one number and halve the other.

Example $3.5 \times 16 = 7 \times 8 = 56$

Distributive rule (partitioning)

The distributive rules for multiplication and division can be written as:

$$a(b \pm c) = ab \pm ac \qquad \frac{b \pm c}{a} = \frac{b}{a} \pm \frac{c}{a}$$

Mental multiplication or division often involves partitioning one of the numbers or expressing it as a difference, and using the distributive rule.

Examples $7 \times 3.9 = 7 \times (4 - 0.1) = 7 \times 4 - 7 \times 0.1 = 28 - 0.7 = 27.3$

$$\frac{250}{6} = \frac{240 + 10}{6} = \frac{240}{6} + \frac{10}{6} = 40 + 1 \text{ rem } 4 = 41 \text{ rem } 4$$

Oral and mental starters

An oral and mental starter can be used to introduce or revise skills of mental calculation or to develop reasoning and proof.

- It can **introduce the main topic**, and many of the teacher-led activities described in this guide can be used in this way (for example, the opening discussion in unit 2).
- It can also be an effective way of **revising skills that are needed for the main topic.**
- Alternatively a starter can be used **to 'keep alive' skills learned earlier that are unrelated to the main lesson.**

Starters can be based on questions in the pupils' book, especially those where items are displayed. Examples are:

p 141 B9   p 157 F4   p 177 F7   p 179 A1   p 221 B3   p 258 H6   p 296 F3

Starter formats

The formats described below have been found very effective and can be adapted to different topics.

Show me

'Show me' boards (small individual whiteboards) are often useful when you want every pupil to respond to a question, for example when doing mental calculations. They can be used for other starter formats, including *True or false?*, *Odd one out* and *Matching*.

How do you do it?

Write a calculation on the board, for example '240 ÷ 15' and ask for as many different (mental or written) ways as possible of doing it. Compare methods.

Today's number is …

Write a number on the board (or get a pupil to choose one). Pupils then make up calculations with that number as the answer or make other true statements about the number. *Today's shape is …* is also possible, and even *Today's algebraic expression is …*

Array

Write a list of numbers or a grid, for example a 4 by 4 grid of mixed positive and negative numbers. Ask questions, such as 'Can you find a pair of numbers whose difference is 5?', 'What is the difference between the total of the first row and the total of the second row?', etc.

The items in the list or grid could also be algebraic expressions.

Display

Display a diagram, graph, calculation, etc. (an OHP transparency is often ideal). For example, display a statistical graph and ask pupils to make statements about what it shows. Or display the calculation 43 × 57 = 2451 and ask what other calculations can be derived from it.

Spider diagram

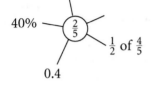

A number or expression is written in the centre of the board. Pupils come and draw lines leading from it to other expressions or words related to it.

True or false?

Read out statements and ask pupils whether they are true or false. Discuss how they could convince someone.

Odd one out

Give sets of numbers, algebraic expressions, words or shapes, and ask for the odd one out, together with the reason why. The odd one out can be different according to the reason given. For example, with 6, 7, 8, 9, 10 the odd one out could be 7 (the only prime number), or 10 (the only multiple of 5).

Matching

Give two sets of numbers or expressions and ask for them to be matched into pairs or larger groups (of equivalent fractions or expressions, for example).

Venn diagram, Carroll diagram

Draw a Venn diagram or Carroll diagram (two-way classification table). Give numbers or shapes to be placed in the correct compartment.

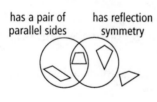
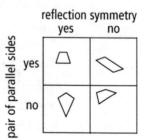

Target number

Write four digits on the board. Pupils have to use them with other signs to make a calculation with a given answer. For example, use 2, 3, 4, 5 to make 44; one way is 2 × 5 + 34.

Definitions

Write some mathematical words on the board and ask for definitions, for example 'rhombus', 'alternate angles'. Discuss whether the definitions offered are correct and clear.

Loop cards or dominoes

Prepare a set of cards (which can be re-used by other classes): each card has a question on it, together with the answer to a question on another of the cards. Make sure no answer appears more than once, so the complete set forms a single loop.

Give out the cards. One pupil reads out their question. The pupil with the answer responds and then reads out their question, and so on.

To avoid leaving anybody out, make sure that there are at least as many cards as pupils. If there are more, then some pupils can have two cards.

Topics for starters

Below, arranged by broad topic area, are some suggestions for other ways these starter formats can be used.

Square, cube, triangle numbers

Array Show some integers that include square, cube and triangle numbers; pupils pick out the type you specify.

Fractions and decimals

Array Show a list of fractions and/or decimals. Pupils put pairs in order of size, find the fraction or decimal halfway between a given pair, order the whole list, etc.

Array/matching Use a mixture of fractions and decimals. Pupils find equivalent pairs.

Today's number is … Write a fraction, e.g. $\frac{3}{8}$. Pupils make up calculations with that as the answer.

Spider diagram Put, say, 0.24 in the centre. Pupils make multiplications with that as the answer.

Array Show a set of decimals. Which pair added, subtracted, multiplied or divided will give the largest or smallest answer? (See page 115 of the pupils' book for division.)

Loop cards Use decimal calculations to be done mentally, e.g. 4.5×6, $1 - 0.37$.

Percentage

Matching Show mixed fractions, decimals, percentages. Pupils find equivalent pairs or threes.

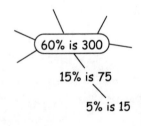

Spider diagram Put a statement in the centre. Pupils add deductions from it, and further deductions from what they have deduced, as shown on the left.

Multiples and factors

Array Show an appropriate set of numbers. Pupils complete statements '… is a multiple of …', '… is a factor of …' in as many ways as possible.

Order of operations

Display A numerical expression is given, for example $8 + 10 \div 5 - 3$. Pupils insert brackets to get a given answer, for example 13.

Rough estimates

Array Show a set of numbers. Pupils say which is closest to the answer to various calculations. (See, for example, the pupils' book page 108, E4.)

Ratio

Spider diagram Write a statement in the centre, such as $A : B = 2 : 3$. Pupils add equivalent statements, such as 'A is $\frac{2}{3}$ of B'.

Spot the rule

2	6	5	9
5	15	11	21
7	21	15	29
9	27	19	37

Activities like those in unit 19 can be done as starters any time after the unit. A variation is as follows.

Array Show a grid of numbers with rules connecting the columns. Find the rule connecting the 1st and 2nd columns, the 1st and 3rd columns, etc.

Think of a number

Show me 'Think of a number' puzzles can be used any time after unit 32.

Substitution into expressions and formulas

Array Show a set of expressions involving letters a, b, c. Pupils work out values for given values of a, b, c.

Linear sequences

Array Show a grid of numbers with linear sequences across and down (and therefore diagonally as well). Pupils give the formula for the nth term of the various sequences in the grid.

Visualising, describing and sketching shapes

Show me Give the name of a 2-D shape and ask pupils to sketch it on their 'show me' boards. Alternatively ask for a piece of information about a 2-D or 3-D shape, for example 'Imagine a square-based pyramid; how many edges does it have?'

Area and perimeter of a rectangle

Show me Ask for the area and perimeter given the dimensions; or ask for the dimensions given the area and perimeter, for example area $20 \, \text{cm}^2$ and perimeter $18 \, \text{cm}^2$.

Estimating angles

Show me Have a competition to get the closest estimate of the size of a given angle (including acute, obtuse and reflex angles).

Angle relationships

Show me Draw diagrams on the board with some given and some unknown angles. Pupils calculate mentally each unknown angle.

Mean, median and range

Array Show numbers in 4 rows and 5 columns. Pupils find the mean, median and range of each row and each column.

① First bites

This is a collection of activities suitable for use in the first one or two weeks of year 7. Their purposes are

- to give pupils of all abilities an enjoyable and confident start to mathematics in the secondary school
- to give you a chance to get to know the pupils and how they work
- to help establish classroom routines and ways of working (whole class, group, individual)
- to give opportunities for homework

They do not not need a high level of number skill, so should be widely accessible.

There may not be time to do all the First bites activities at the beginning of year 7. Some may be left for later, perhaps as starting points for related units of work.

p 4	**A** Spot the mistake
p 4	**B** Four digits
p 4	**C** 4U + 1T
p 5	**D** Finding your way
p 6	**E** Gridlock
p 8	**F** Patterns from a hexagon
p 11	**G** Shapes on a dotty square
p 12	**H** Find it!

Essential	**Optional**
Sheets 45, 46 and 78	Sheets 47, 48, 49 (blank grids) and 58
Up to 15 dice for a class of 30	Square and triangular dotty paper (sheets 1 and 2),
Sharp pencils, pairs of compasses, rulers,	3 by 3 pinboards, rubber bands, tracing paper,
coloured pencils, board compasses	OHP transparency of square dotty paper
Angle measurers	

A Spot the mistake (p 4)

T

These two resource sheets give pupils of all abilities an opportunity, in a light-hearted context, to spot some mathematical errors – and some non-mathematical ones! It is more fun for pupils to work in pairs, rather than on their own.

Sheets 45 and 46

B Four digits (p 4)

T

This activity will tell you something about pupils' knowledge of number operations and symbols (for example, brackets) and their arithmetic skills. It also gives an opportunity for co-operative group work.

◊ Decide together on the four digits to be used. They do not have to be all different. 0 is not very helpful.

◊ You can allow free rein at first as far as the rules are concerned. Pupils will probably come up with some ground rules themselves, and then you can establish rules for everyone, for example:

- All four digits must be used.
- No digit can be repeated unless it occurs twice in the set.
- Digits can be used in any order.
- Any operations can be used. (Brackets may be needed and √ may be suggested by pupils.)
- Digits can be combined to make two- and three-digit numbers.
- Results must be whole numbers (for example, $43 \div (2 + 1) \neq 14$).

◊ You could start by asking for ways to make, for example, 10.

◊ Pupils could work in groups, each group making a collection. An element of competition could be introduced. Alternatively, groups could be given ranges of numbers (1–20, 21–40, …).

◊ It is necessary to record the completed numbers and the methods used to arrive at them, for example, a list:

1	11	21
2	12	22
3	13	etc.
4	14	
5	15	
6 $1 + 3 + 4 - 2$	16	
7	17	
8 $2 + 3 + 4 - 1$	18	
9	19 $12 + 3 + 4$	
10 $1 + 2 + 3 + 4$	20	

◊ In one school the results were recorded on a large chart and put on the wall. This was added to over the year. (It works particularly well if there is some reward for completing gaps – merits, credits, etc.)

◊ Calculators may be used if necessary, although many pupils should do well without them.

◊ Pupils may realise that results often come in pairs, for example:
$$23 - 14 = 9, \quad 23 + 14 = 37$$
then with some swapping around:
$$32 - 14 = 18, \quad 32 + 14 = 46, \quad \text{etc.}$$

Follow-up

Each pupil can choose their own set of four numbers. Some sets of numbers (for example, 6, 7, 8, 9) are more difficult than others and may lead to demotivation.

ℂ **4U + 1T** (p 4)

◊ Start by asking for a two-digit number. (If 13, 26 or 39 are suggested, find a sneaky way of avoiding them!) Write the number on the board. Say that you are going to multiply the units digit by 4 and then add on the tens digit (so 37 becomes $4 \times 7 + 3$, giving 31). Write the result on the board.

Ask pupils to do the same to this new number. Write the result. Repeat a few times until all pupils have understood the rule for generating the next number. Do not go on too long or you will form a loop. It is best for pupils to discover the loop for themselves.

You may come across a single-digit number in the process. If not, you should introduce one and discuss how to deal with it.

Ask pupils each to choose a two-digit number of their own, use the rule to make a chain of numbers and watch what happens as their chain grows.

After a while, someone will notice they have come back to a number they had before. When this happens, discuss it with the class. Pupils should realise that once a loop is formed, no new numbers are generated, but they do not always find this obvious!

◊ Questions for investigation are:
   • Will all numbers form loops?
   • How long are the loops?
   • How many different loops are there?
   • Do any numbers go straight to themselves? (Yes: 13, 26 and 39!)

There is nothing special about 4 as the multiplier: it just gives short chains. Pupils can investigate other multipliers. How many chains are there in each case? (The rule 2U + 1T produces a nice overall diagram with every number connected to one big loop, except for the multiples of 19.)

Pupils can be challenged to work backwards. For example, if the rule is 2U + 1T, what numbers generate 15 as their next number?

D Finding your way (p 5)

This gives practice in using left and right and in reading a simple map.

◊ Before discussing the picture and the questions, you could use a plan of your school. Give pupils a list of instructions from the classroom to somewhere else and ask them to guess the destination. They can then make up instructions for one another.

You could ask pupils to shut their eyes and imagine where they are going as you give them instructions.

Similar work can be based on local maps.

E Gridlock (p 6)

This game gives you an opportunity to find out pupils' addition (and subtraction) skills. Pupils can also develop and explain strategies to win.

Up to 15 dice for a class of 30
Optional: Sheets 47, 48 and 49 (blank grids)

◊ To help them understand the scoring system, pupils could complete the grids on sheet 47 and work out the scores. Some schools have used this sheet for homework.

'I almost didn't use sheet 47 but it turned out to be most useful.'

◊ Initially the class could play together, with you rolling the dice and calling the numbers. Then the game can be played in groups of two or more.

◊ More able pupils may think that the game is a trivial exercise requiring only simple addition skills. Emphasise early on that they should be thinking about good strategies to maximise their chance of winning.

To play 'Gridlock'

Each pupil draws a square grid (start with 3 by 3 grids) and marks off the top left-hand section as shown.

The caller rolls a dice and calls out the number. Each pupil writes the number in any empty square in the section shown shaded on the right.

4	2	
3	3	

Repeat until each square in that section is filled.

Each number must be written in the grid before the next is called and a number can't be changed once it is written.

Each pupil adds up their numbers in the rows, columns and diagonal and writes the totals in the empty squares as shown on the right.

4	2	6
3	3	6
7	5	7

Each pupil adds up their points.

• Score 2 points for a total that appears twice.
• Score 3 points for a total that appears three times.
• Score 4 points for a total that appears four times ... and so on.

The grid above scores 4 points (6 and 7 both appear twice as a total).

After a number of rounds (decided by you), pupils add up their points and the one with most points is the winner.

◊ After playing on 3 by 3 grids, play the game on larger ones.

◊ Once pupils have played the game a few times, ask them to describe any strategies they use in placing the numbers on their grids. For example, if a number is rolled twice it is better to place the numbers diagonally,

'I tried moving on to 5 by 5 grids and it was useful. I rolled 16 dice, wrote the numbers on the board and challenged pupils to achieve 9 points and 0 points. This exposed the value of the diagonal symmetry ... Pupils greatly enjoyed this and preferred this variation to the original game.'

for example

5	
	5

or

	5
5	

rather than

5	5

◊ Now you can alter the rules as follows. First, the numbers called out are written at the side of the grid. When all numbers have been called, they are then placed in the grid. Pupils can think about how to get the maximum possible score with a particular set of numbers.

◊ One variation is for the winner to be the person with the fewest points. Pupils can discuss how their winning strategies change in this case.

Another variation is to use two dice to generate larger numbers.

Follow-up

In E1 to E11, the later questions are more difficult.

Remind pupils that they can only use numbers on an ordinary dice (1 to 6) to solve these problems.

E1 In (b), emphasise that their problems should be solvable without any guesswork or mind reading! Encourage more able pupils to make up problems that give the minimum necessary information.

This could be set as a homework task.

◊ You could ask pupils how changing the system of scoring points would affect the game, for example:

- score 2 points for a total that appears twice
- score 4 points for a total that appears three times
- score 6 points for a total that appears four times ... and so on

◊ A different version of the game is for pupils to cross out any totals that repeat and to add the remaining totals to give their score for that round. The winner could be the person with the most or fewest points.

𝔽 Patterns from a hexagon (p 8)

This work is to help pupils develop skills with compasses and rulers that are needed later to construct triangles and angles. Pupils also analyse patterns and make decisions about how to construct them.

At the start of the year many pupils have coloured pencils and geometry sets, so capitalise on this.

> Sharp pencils, pairs of compasses, rulers, coloured pencils, board compasses

T

'I discovered that only about half the class had used compasses before.'

◊ Many pupils find it difficult to draw a circle with a pair of compasses. They may need to draw circles and simple patterns before they feel confident enough to try the more difficult patterns.

Many pupils will find it helpful to see a demonstration of how to draw a regular hexagon. They must be able to draw a regular hexagon in order to draw the patterns on page 9.

Common problems include:

- not realising that the point of the compasses is moved to the point where the last arc crosses the circumference (and not at the end of the arc) for subsequent arcs to be drawn
- not realising that, to draw the hexagon, you join points where the arcs cross the circumference (and not the ends of the arcs)

Although pupils may have drawn them before, it may be helpful to demonstrate on the board or OHP how to draw the seven-circle or petal designs shown below.

◊ This work provides good material for wall displays. In one school, the hexagon designs were used to make mobiles.

◊ You may want pupils to leave construction lines so you can check their methods.

◊ The construction of the patterns on page 9 offers more of a challenge, and the construction gets more involved further down the page.

The last two designs can be drawn as follows:

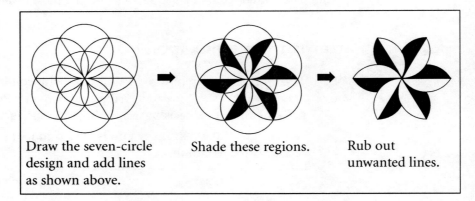

Draw the seven-circle design and add lines as shown above.

Shade these regions.

Rub out unwanted lines.

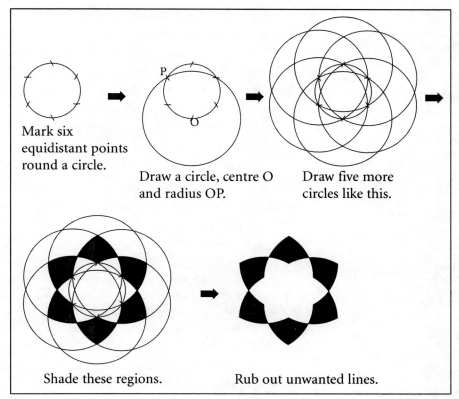

Mark six equidistant points round a circle.

Draw a circle, centre O and radius OP.

Draw five more circles like this.

Shade these regions.

Rub out unwanted lines.

Follow-up

If they know about symmetry, pupils could try to draw a pattern with
0 lines of symmetry, 1 line of symmetry etc.

Ⓖ **Shapes on a dotty square** (p 11)

Pupils create shapes and use mathematical language to describe them.
They can also decide on their own lines of investigation.

> Optional: Square and triangular dotty paper (sheets 1 and 2), sheet 58,
> 3 by 3 pinboards, rubber bands, tracing paper, OHP transparency of
> square dotty paper

◊ Square dotty paper can be used or pupils can draw grids of dots on square
  paper. Sheet 58 has the grids already ruled off.

◊ Establish rules for drawing shapes on the pinboard/grid.
  • Only the 9 pins/dots can be used.
  • All corners must be at a pin/dot.
  • The types of shape shown in the pupil's book are disallowed (ones with
    'crossovers' or 'sticking out' lines).

*'For some pupils,
drawing out the grid
was the hardest part.'*

◊ Ask pupils to draw a few different shapes following the rules above.

There is likely to be some discussion on the possible meanings of 'different' and 'same' here. 'Different' is usually taken to mean non-congruent. Tracing paper helps pupils identify shapes that are the same.

Look at some of their shapes together.

• What properties have they got?
   (Number of sides, angles, symmetry, parallel sides, area, ...)

• Do pupils know names for any of the shapes?
   (Triangle, quadrilateral, rectangle, parallelogram, hexagon, ...)

◊ There are various ways to structure this activity. Some trial schools generated a collection of questions from which pupils chose. For example:

   What shapes have the most sides?
   How many different squares?
   How many different triangles?
   How many shapes have reflection symmetry?
   How many shapes have rotation symmetry?
   What different areas can you make?
   How many ways can you put the same triangle on the grid?
   What shapes can be made with 1, 2, 3, ... right angles?

Pupils can choose a question (or pose one of their own) and write up their solution. These could then be displayed.

For example, the 23 different polygons with reflection symmetry are:

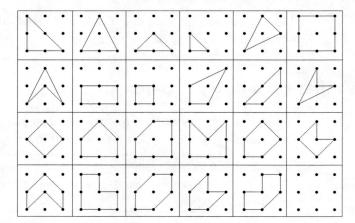

◊ An alternative structure is to begin by asking how many different triangles can be found. The 8 different triangles are:

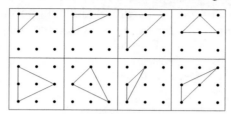

Now discuss the properties of the triangles. For example:

Which has the greatest area?
Which has a right angle?
Which has reflection symmetry?
Which are isosceles?

Pupils can now consider the different quadrilaterals that can be found and their properties. The 16 different quadrilaterals are:

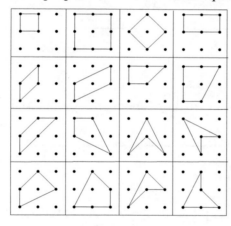

Pentagons, hexagons and heptagons can be considered but the number of different shapes may be rather daunting!

The numbers of different polygons of each type are:

Number of sides	Number of different polygons
3	8
4	16
5	23
6	22
7	5

This gives 74 different polygons.

'I put the class into teams, following a lesson and a homework, to find all possible polygons, to collate results and present.'

◊ One extension is to consider polygons on a different grid. Suggestions appear in the pupil's book.

The hexagonal 7-pin grid yields 19 different polygons which is a manageable number for most pupils to find. The polygons are:

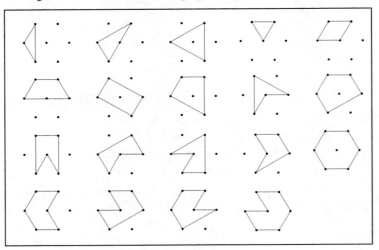

⊞ **Find it!** (p 12)

Pupils follow instructions about angles and distances to locate points on a map.

> Sheet 78
> Angle measurers

◊ Make sure that pupils understand about turning through an angle from the direction in which they are initially facing. You could demonstrate this by walking forward holding a ruler out in front of you to show the direction in which you are facing. When you stop and turn, you move the ruler through an angle.

Now draw your path and the angle of turn on the board (as in the diagram in the pupil's book).

Pupils with experience of LOGO may be familiar with this use of angles.

A Spot the mistake (p 4)

1 Off to Benidorm in June

Fish tank framework is an impossible object.

Table left-hand rear leg is longer than the others.

Vase on the table is an impossible object.

Sofa is an impossible object.

Socket on left-hand wall has holes in wrong orientation.

Mirror reflection is incorrect; 'TAXI' and the clock face are the wrong way round.

Front door has handle and hinges on the same side.

The '33' above the front door is the wrong way round.

Vacuum cleaner plug has only one pin.

Vacuum cleaner hose has an extra hose tangled in it.

Triangles on shelf: right-hand one is an impossible object.

View of window blind is impossible.

2 Taxi to the airport

Clock has a back-to-front 3, and 8 where it should be 9.

P (parking) sign is on a road with double yellow lines.

Stop sign on road – S is wrong way round.

The word STOP (and the road marking) is on the wrong side of the road.

Bicycle has no front wheel.

Airport sign says 5 cm (centimetres).

Traffic lights have a right turn arrow to a no-entry street.

Rollerblader has one ice skate on.

Right-hand no-entry sign is pointing upwards.

Pedestrian crossing markings on the road should be rectangles.

Pedestrian crossing beacon is the wrong shape.

Street lamp on top of the no-entry sign is facing up.

A vegetarian butcher would not do much business!

Low bridge sign says min(imum) and should say max(imum).

Taxi is going to the airport, so should have turned left.

('Tax to rise 150%' is not necessarily wrong – it could!)

3 At the airport

A plane is flying upside down.

The plane taking off has no tail wings.

Passengers are walking along the wing of the waiting plane.

Waiting plane has RAF insignia on tail.

Waiting plane has a ski instead of a wheel.

Wind socks are blowing in opposite directions.

Tannoy message says 'train' instead of 'plane'.

Tannoy message contradicts time on clock.

Christmas tree contradicts Easter eggs sign.

Tax free sign: you cannot save 200%.

Suitcases are ticketed to Rome, and flight is to Benidorm.

Suitcase on weight machine has a square wheel.

Luggage weight sign says WAIT, not WEIGHT.

Luggage weight is in g(rams) and should be kg (kilograms).

The 'All departures' sign points to a no entry corridor.

You cannot 'Ski the Pyramids'.

You cannot 'Ice skate the Amazon'.

4 The hotel reception

Clock reads 14:60, and the minutes must be less than 60.

Sign above toilet doors says 'Welcome to BeMidorN.'

Calendar on reception desk says 31 June (only 30 days in June).

A Christmas tree in June is rare!

Toilet door pictures are the wrong way round.

Right-hand toilet door has handle and hinges on the same side.

Change sign – 100 pts should be 1000 pts, and the peseta has of course been replaced by the euro.

Double rooms are cheaper than single rooms.

Atlantic views should read Mediterranean views in Benidorm.

Left-hand rear leg on table is longer than other three.

Tickets on luggage have changed since the airport to Roma and Home.

Plant doesn't sit in its flower pot.

Lift is on ground floor (the lowest in the list above the door) but shows on the list (and the call buttons) as going further down.

List of floors above lift door is missing floor 4.

Sign in lift mirror – S is wrong in reflection.

Sign in lift has a weight limit that is silly. (The vase on the table is not an impossible object!)

5 By the pool

You don't get whales in the Mediterranean.

The flag at the top of the boat's mast should be blowing forwards.

Speed boat and water skier are not connected.

Plane is flying backwards if it is pulling the banner.

Banner should read 24 hours not 26.

Weather vane NSEW are wrong.

If the time is 23:30 the sun would not be out.

Temperature of ⁻26°C is a bit chilly for sunbathing.

Water level in man's jug should be horizontal.

Sun lounger nearest to front is missing a leg.

Gazebo on the left-hand corner of the balcony is an impossible object.

Diving board heights are in millimetres, and should be in metres.

Lower diving board is at a greater height (8) than the upper (4).

There appears no means of access to the lower diving board.

'Do not feed fish' sign is unlikely in a swimming pool.

Man fishing not possible (we hope) in a swimming pool.

Swimming pool depth signs are wrong.

Shark in swimming pool.

Stairs and railings up to balcony create an impossible object.

Sangria jugs of 75 l(itres) would be a bit big!

Children shown standing at the left-hand side of the pool where depth is 5 m(etres).

Depth shown as 10 cm where the diving boards are.

D Finding your way (p 5)

D1 Robin Hall

D2 The pupil's journey

E Gridlock (p 6)

E1 (a)

2	3	5
3	4	7
5	7	6

Points scored: **4**

4	1	5
3	6	9
7	7	10

Points scored: **2**

1	6	5	12
5	3	1	9
6	4	6	16
12	13	12	10

Points scored: **3**

(b) The pupil's problems

E2 Examples of grids that score 2 points are:

5	3	8
6	2	8
11	5	7

3	6	9
5	2	7
8	8	5

E3 Examples are:

(a)

6	6	12
4	3	7
10	9	9

(b)

4	6	10
6	3	9
10	9	7

E4 (a) 2 points (b) 0 points

E5 (a)

4	**2**	6
2	**1**	3
6	3	5

(b)

3	**1**	4
2	**5**	7
5	6	8

(c)

5	**2**	**3**	10
4	6	**4**	14
1	**1**	**1**	3
10	9	8	12

E6 Examples are: 1, 2, 3 and 5; 2, 3, 4 and 6.

E7 The pupil's explanation

E8 The pupil's explanation

E9 (a)

6	**4**	10
4	**3**	7
10	7	**9**

(b)

1	**5**	**6**
4	**6**	10
5	11	7

E10 Examples are:

(a)

1	2	6	9
2	3	5	10
6	5	4	15
9	10	15	8

(b)

1	2	5	8
2	3	6	11
5	6	4	15
8	11	15	8

(c)

5	1	2	8
5	2	3	10
4	6	6	16
14	9	11	13

E11

1	5	5	11
1	3	4	8
6	5	6	17
8	13	15	10

Sheet 47

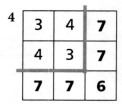

1

6	5	**11**
1	1	**2**
7	**6**	**7**

Points scored: **2**

2

3	4	**7**
6	5	**11**
9	**9**	**8**

Points scored: **2**

3

3	5	**8**
2	4	**6**
5	**9**	**7**

Points scored: **0**

4

3	4	**7**
4	3	**7**
7	**7**	**6**

Points scored: **4**

5

6	1	2	**9**
3	4	6	**13**
1	5	1	**7**
10	**10**	**9**	**11**

Points scored: **4**

6

5	3	5	**13**
3	1	4	**8**
2	6	1	**9**
10	**10**	**10**	**7**

Points scored: **3**

Ⱶ **Find it** (p 12)

The spy camera is at (11, 4)

The x-ray binoculars are at (13, 9)

The coding kit is at (5, 14)

The radio is at (24, 13)

Symmetry

This involves reflection and rotation symmetry.

There is more material here than one pupil could be expected to do in the time likely to be available. Once you have seen how much pupils already know about symmetry, you can choose appropriate work.

T

| p 13 | **A** What is symmetrical about these shapes? | Recognising reflection and rotation symmetry |

| p 14 | **B** Reflection symmetry | Recognising lines of symmetry Drawing reflections |

T

| p 15 | **C** Rotation symmetry | Order of rotation symmetry |

| p 16 | **D** Making designs | Drawing designs with rotation symmetry |

| p 18 | **E** Rotation and reflection symmetry | |

| p 19 | **F** Times and dates | Symmetry in groups of digits |

| p 20 | **G** Symmetry tiles | Making symmetrical patterns with tiles |

| p 23 | **H** Pentominoes | Symmetry in pentominoes |

Essential	**Optional**
Mirrors	OHP transparency made from sheet 116
Scissors	
Tracing paper	
Square dotty paper (sheet 1)	
Triangular dotty paper (sheet 2)	
Sheets 60, 62, 64, 65, 117 and 120–122	
Practice booklet pages 3 to 8	

Ⓐ **What is symmetrical about these shapes?** (p 13)

Discussion should show how much pupils know already about reflection and rotation symmetry.

Optional: OHP transparency made from sheet 116, tracing paper

◊ One way to generate discussion is for pupils to study the page individually, then discuss it in small groups; then you can bring the whole class together and ask for contributions from the groups.

◊ Most pupils should be able to describe the reflection symmetry of the shapes. Some may realise that shapes with only rotation symmetry (B, C, D, I) are symmetrical in some way but be unable to describe how. Others may know about rotation symmetry already.

B **Reflection symmetry** (p 14)

Pupils recognise lines of symmetry and draw reflections across vertical, horizontal and sloping lines.

Mirrors, sheets 60 and 62
Optional: Scissors

B1 Many pupils think the diagonals of a rectangle and a parallelogram are lines of symmetry. B1 offers the opportunity to deal with this misconception.

C **Rotation symmetry** (p 15)

Tracing paper, sheet 117
Optional: Square dotty paper

◊ The shape on the page is the first one on sheet 117. Pupils can make and trace their own copy of this shape on square dotty paper or trace the one on the sheet.

◊ Emphasise that the shape can be rotated by putting a pencil point at the centre of rotation and turning the tracing round this fixed point.

◊ If a shape has rotation symmetry of order 1, then every point in its plane is a 'centre of rotation symmetry', because if the shape is rotated through 360° about any point it returns to its original position. Rotation symmetry of order 1 is of no interest and is ignored after question B1.

D **Making designs** (p 16)

Tracing paper, square dotty paper, triangular dotty paper,
sheets 120 and 121

E Rotation and reflection symmetry (p 18)

> Sheet 122, square dotty paper

E4 Pupils can extend this by finding all possible ways to shade four squares to produce a design with rotation symmetry.

F Times and dates (p 19)

Pupils find lines of symmetry in groups of digits (times and dates) where two lines of symmetry are possible.

> Mirrors

◊ Many pupils find it harder to identify a horizontal line of symmetry than a vertical one. It may be worth leading a brief discussion after F1 has been completed.

'Surprisingly, pupils were not proficient at putting dates into figures.'

◊ Some pupils may not be familiar with the method of writing dates used in F2 onwards. It may be beneficial to discuss this with pupils before they begin this work.

◊ Emphasise that dates found in F5 and F6 have to be possible. For example, 83:38:83 is not a possible date!

G Symmetry tiles (p 20)

Pupils consolidate work on reflection symmetry by using tiles to make symmetrical patterns. The tiles are also used to play a game.

> Scissors, mirrors, sheet 64 (copied on card if possible),
> sheet 65 (one for each group of players)

◊ In problems G1 to G6 there is no need for pupils to draw diagrams to show their results but some may wish to do so.

Symmetry tiles game

◊ The game should be self-correcting: hopefully, players will protest at an invalid move. Some care is needed in organising the groups: each group should have one pupil who is confident enough about symmetry to recognise invalid moves. Although the game can be played with four players, having only two or three makes it faster and more enjoyable.

◊ You may have to clarify one or two things: pupils can put their cards down on either side of the dotted line; they don't have to complete the symmetry at every move (though they could play that way).

◊ When using rotation symmetry, pupils should ignore the dotted lines.

◊ You could challenge pupils to complete board 3 with two lines of symmetry and rotation symmetry order 2, or rotation symmetry order 4 but no line symmetry.

Ⓗ Pentominoes (p 23)

Square dotty paper

Ⓑ Reflection symmetry (p 14)

B1 (a) Yes   (b) No   (c) Yes
(d) Yes   (e) Yes   (f) No

B2 Pupil's completed diagrams on sheets 60 and 62.

B3 (a) (i) % $ ✳ ∧ ⊗ ∈

        (ii)

        (iii)

(b) ✳ ⊗ ∈ They have a horizontal line of symmetry.

(c) ✳ ∧ ⊗ They have a vertical line of symmetry.

(d) % $ They have rotation symmetry of order 2.

(e) ⊗ It has four lines of symmetry and rotation symmetry of order 4.

Ⓒ Rotation symmetry (p 15)

C1 Centres of rotation marked on sheet 117

Shapes with rotation symmetry	Order of rotation symmetry
A	4
B	3
C	2
E	4
F	2
G	3
I	2
J	6
K	8
L	2
M	4

Ⓓ Making designs (p 16)

D1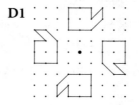

D2 Completed designs on sheet 120

D3 (a) 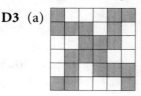   (b)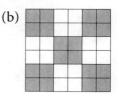

(c)

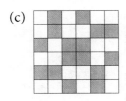

D4

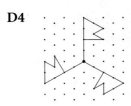

D5 Pupil's completed designs on sheet 121

D6 (a)   (b)

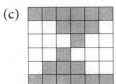

(c)

D7 (a)   (b)   (c)

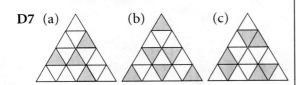

E **Rotation and reflection symmetry** (p 18)

E1

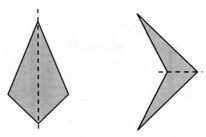

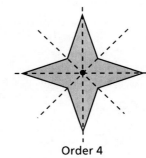

Order 4

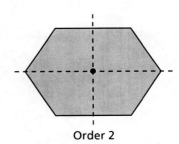

Order 2

E2 (a) Rotation symmetry of order 2
No lines of symmetry

(b) Rotation symmetry of order 3
Three lines of symmetry

(c) No rotation symmetry
No lines of symmetry

(d) No rotation symmetry
One line of symmetry

(e) Rotation symmetry of order 3
No lines of symmetry

(f) Rotation symmetry of order 2
Two lines of symmetry

(g) Rotation symmetry of order 2
No lines of symmetry

(h) Rotation symmetry of order 2
Two lines of symmetry

(i) No rotation symmetry
One line of symmetry

(j) No rotation symmetry
No lines of symmetry

(k) Rotation symmetry of order 2
No lines of symmetry

(l) No rotation symmetry
No lines of symmetry

(m) No rotation symmetry
No lines of symmetry

(n) Rotation symmetry of order 2
No lines of symmetry

E3 (a) Order 2

(b) Yes, two lines of symmetry

E4 There are sixteen ways to shade four squares to make a pattern with rotation symmetry (plus twelve that are 90° rotations of some of the sixteen). Pupils have to find eight different ways.

The sixteen ways are shown below.

Rotation symmetry of order 2

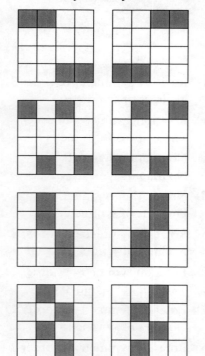

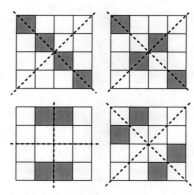

Rotation symmetry of order 4

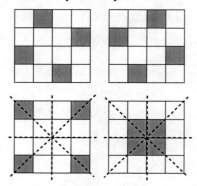

E5 (a) The pupil's pattern from

(b) The pupil's pattern from

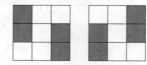

or 90° rotations of these

(c) Some examples are

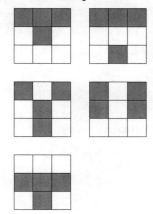

(d) Some examples are

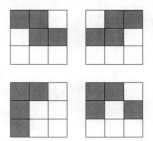

F **Times and dates** (p 19)

F1 (a) 01:18, 03:38, 11:18, 01:00, 13:13, 01:10 and 10:01 are symmetrical.

(b) 01:10 and 10:01 have two lines of symmetry.

(c) The pupil's three times with one line of symmetry

(d) One time from 11:11 and 00:00

(e) 01:10 and 10:01

F2 Two lines of symmetry

F3 (a) None

(b) Two lines, order 2

(c) No lines, order 2

(d) One line

F4 (a) 04:02:33 none

(b) 31:10:81 one line

(c) 08:11:80 two lines, rotation order 2

(d) 09:11:60 no lines, rotation order 2

F5 The pupil's two dates with two lines of symmetry

F6 The pupil's date with no lines of symmetry, but rotational symmetry order 2.

G **Symmetry tiles** (p 20)

G1, G2 and **G3** The pupil's patterns

G4

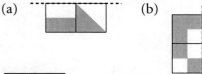

G5 In each case, there is another solution with the tiles on the other side of the dotted line.

(a)   (b)

G6

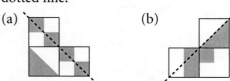

G7 In each case, there is another solution with the tiles on the other side of the dotted line.

(a)   (b)

G8 In each case, there is another solution with the tiles on the other side of the dotted line.

(a)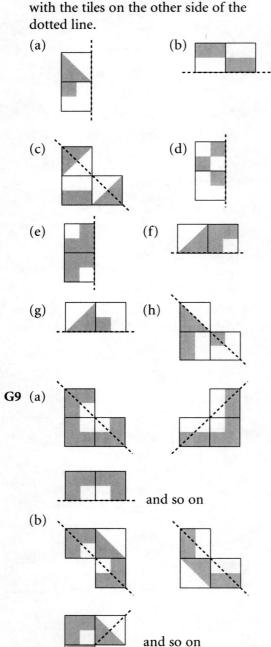

(b)

(c)

(d)

(e)

(f)

(g)

(h)

G9 (a)

and so on

(b)

and so on

(c)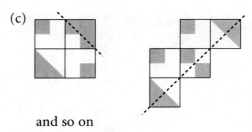

and so on

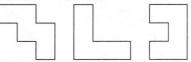**Pentominoes** (p 23)

H1 The pupil's pentomino from

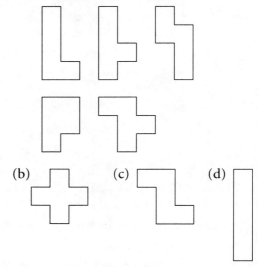

H2 (a) The pupil's pentomino from

(b)　　　　(c)　　　　(d)

H3 (a) (i) The pupil's shape; examples are

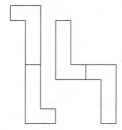

(ii) The pupil's shape; examples are

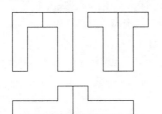

(iii) The pupil's shape; examples are

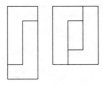

(b) The pupil's design; examples are

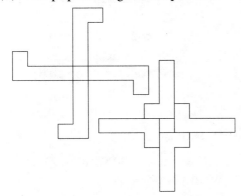

H4 The pupil's designs with
- (a) reflection symmetry but no rotation symmetry
- (b) rotation symmetry but no reflection symmetry
- (c) reflection symmetry and rotation symmetry

What progress have you made? (p 24)

1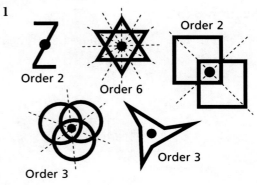

2 (a)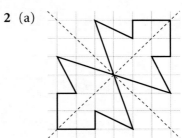

(b)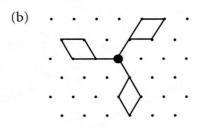

3 The pupil's pattern; examples are

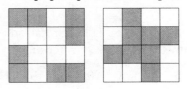

4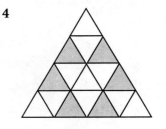

Practice booklet

Section B (p 3)

1 (a) Lines 1, 2, 3, 4 and 5 (all of them)

(b) Neither is a line of symmetry.

(c) Lines 1 and 5

2 (a) Four ways

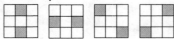

(b) Four ways

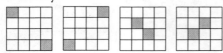

(c) Eight ways

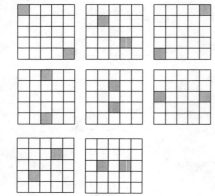

Section C (p 4)

1 Designs A, C and D have rotation symmetry (of order greater than 1).

2 (a) 5   (b) 8   (c) 2   (d) 4
  (e) 3   (f) 3   (g) 2   (h) 2

Section D (p 5)

1 The pupil's completed designs with rotation symmetry of order 4

2 The pupil's completed designs with rotation symmetry of order 2

3 The pupil's completed designs with rotation symmetry of order 3

Section E (p 6)

1 (a) No lines of symmetry
Rotation symmetry of order 3

(b) Two lines of symmetry
Rotation symmetry of order 2

(c) One line of symmetry
No rotation symmetry

(d) No lines of symmetry
No rotation symmetry

(e) Two lines of symmetry
Rotation symmetry of order 2

(f) No lines of symmetry
Rotation symmetry of order 3

2 (a) No lines of symmetry
Rotation symmetry of order 2

(b) One line of symmetry
No rotation symmetry

(c) No lines of symmetry
No rotation symmetry

(d) No lines of symmetry
Rotation symmetry of order 2

Section F (p 7)

1 (a) 1331, 1301, 1811, 1380

(b) 1881

(c) 1961, 1881, 1691

2 1800, 1801, 1803, 1808, 1810, 1811, 1813, 1818, 1830, 1831, 1833, 1838, 1880, 1881, 1883, 1888, 1961

3 Examples include
18 + 13 = 31, 81 + 30 = 111

4 Examples include
13 × 10 = 130, 88 × 101 = 8888

5 (a) H, I, N, O, S, X, Z

(b) Examples include
ONNO, ISSI, SHHS

Section H (p 8)

1

Rotation symmetry of order 4 and four lines of symmetry

No rotation symmetry or reflection symmetry

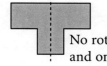

No rotation symmetry and one line of symmetry

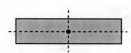

Rotation symmetry of order 2 and two lines of symmetry

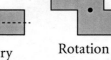

Rotation symmetry of order 2 and no reflection symmetry

2 It has one line of reflection symmetry and no rotation symmetry.

3 Examples are

(a)

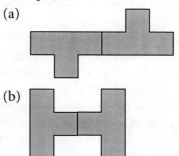

(b)

4 The pupil's designs with

(a) reflection symmetry but no rotation symmetry

(b) reflection symmetry and rotation symmetry

(c) rotation symmetry of order 4

③ Number skills

This unit gives opportunities to practise arithmetical skills with whole numbers, up to long multiplication and division. In many instances these skills are practised in the context of an investigation or problem.

p 25   **A** Investigations and problems

p 27   **B** Multiplying by a two-digit number

p 29   **C** Division and remainders

p 30   **D** Division by a two-digit number

p 31   **E** Problems

p 32   **F** Remainders with a calculator

Essential

Sheet 24

Practice booklet pages 9 to 13

Ⓐ Investigations and problems (p 25)

Number magic

With two-digit numbers pupils may notice that:

- The first answer is always in the 9 times table.
- The final answer is always 99 (treating 9 as 09).

Very high attainers could be helped to prove these results, by using letters to stand for digits.

$$
\begin{array}{cc}
\text{Tens} & \text{Units} \\
A - 1 & B + 10 \\
\cancel{A} & \cancel{B} \\
-\quad B & A \\
\hline
A - 1 - B & B + 10 - A \\
+\quad B + 10 - A & A - 1 - B \\
\hline
9 & 9 \\
\end{array}
$$

With three-digit numbers the final answer is always 1089 (treating 99 as 099).

With four-digit numbers the final answer depends on the middle pair of digits. If the 2nd digit is greater than the 3rd, the result is 10890. If the 2nd and 3rd digits are equal, the result is 10989. If the 2nd digit is less than the third, the result is 9999.

Encourage more able pupils to persist and try to find a pattern in their results.

Largest and smallest

◊ The process always ends up with the number 6174.

Target 1000

◊ A total of 1000 is impossible, but there are very many ways of getting a total of 999. The units column has to add to 19, the tens column to 18 and the hundreds column to 8. For example, 79 + 84 + 516 + 320.

'I set this as homework. Pupils enjoyed it and a lot got their parents to help.'

Fruit punch

◊ A hint is to try values of P first, as P occurs most often. Since the maximum 'carry' is 1, P is one more than A.

45581	45581	56683	56683	56683	56683	56683
5826	5892	6819	6827	6824	6891	6821
51407	51473	63502	63510	63507	63574	63504

Columns

> Sheet 24 (one for every 6 pupils)

◊ Multiplying a number in column 3 by 4 will always give a result in column 2. (The hundred square may need to be extended downwards.) Some pupils may be able to go on to predict similar results for other columns.

◊ One extension is to look at square numbers and which columns they are in (and which they are never in).

Making multiplications with three digits

◊ With three different digits there are 6 different multiplications (counting 34 × 6 the same as 6 × 34)

$$34 \times 6 \quad 43 \times 6 \quad 36 \times 4$$
$$63 \times 4 \quad 46 \times 3 \quad 64 \times 3$$

The largest result is 43 × 6 and the smallest is 46 × 3.

◊ To obtain the largest result, it is clear that the smallest number must go in the units column (to minimise its effect) giving two possibilities: 43 × 6 or 63 × 4. It must be the first case as 6 × 3 is greater than 4 × 3.

In general, choose the largest digit to be the single-digit multiplier and use the next largest in the tens column.

⒝ Multiplying by a two-digit number (p 27)

Pupils consider a variety of methods to multiply by a two-digit number. Those who do not already have a method to multiply by a two-digit number may like to adopt one of these.

◊ You could focus on each method in turn and see if pupils, working in groups, can explain it and demonstrate using it for another problem, say, 23 × 24.

Some teachers have split the class into groups and allocated each group one method to study and report on.

'We have a descendant of Napier in the class, so this was particularly interesting.'

◊ When setting out their grids for method B, pupils need to remember that the numbers to be multiplied are positioned at the top and right-hand side of the grid.

This method is often referred to as the gelosia method.

◊ Method C is in many ways the most illuminating. It can be applied later to expanding brackets in algebra. Some pupils like this method (or method B) so much that they adopt it as their standard method.

◊ Pupils could decide how to use these methods to multiply numbers with more than two digits.

⒞ Division and remainders (p 29)

◊ Initially each pupil could tackle the three problems on their own. Then they can discuss their solutions in groups.

Each group could make up a problem for the whole class to do.

⒟ Division by a two-digit number (p 30)

◊ Pupils may have learned the 'standard' long division algorithm and feel quite happy with it. However, many do not find it straightforward and the other method is a fairly efficient alternative.

E Problems (p 31)

◊ Pupils can try these without a calculator to give further practice in written or mental methods. However, if the problems prove too hard or too tedious without a calculator, pupils could use a calculator for this section.

F Remainders with a calculator (p 32)

◊ The number of people on the fifth coach can be found by doing 0.711... × 52. Alternatively, you can work out 4 × 52 and subtract the result from 245.

Some calculators will not give a whole number when the decimal 0.711... is multiplied by 52. You may need to discuss the limitations of the calculator here. Emphasise that the decimal 0.711... goes on for ever and their calculator cannot store it all, so occasionally this causes problems. Pupils may feel uneasy about this and prefer to use the second method described above.

B Multiplying by a two-digit number (p 27)

B1 (a) 2331
(b) The pupil's check that 37 × 63 is also 2331

B2 (a) 1026   (b) 1708   (c) 3388
(d) 3818   (e) 738   (f) 2997
(g) 4836   (h) 5208

B3 (a) The first line should be 320 (16 × 20).
(b) 384

B4 (a) 3645   (b) 8037   (c) 19874
(d) 38868

B5 (a) With four different digits there are 12 different multiplications (counting 34 × 56 the same as 56 × 34):

34 × 56	43 × 56	34 × 65
43 × 65	36 × 54	63 × 54
36 × 45	63 × 45	46 × 53
64 × 53	46 × 35	64 × 35

The largest result is 63 × 54 and the smallest is 46 × 35.

(b) The largest result is 92 × 75 and the smallest is 59 × 27.

(c) Consider, for example, the digits 3, 4, 5 and 6. For the largest result, the positions of the 6 and the 5 in the tens place are obvious. The choice is then between 64 × 53 and 63 × 54. In the second case the larger units figure multiplies the larger tens figure, so this gives the larger result. (Method C on page 27 of the pupil's book makes this clear.)

B6 The largest result is with 752×84. The positions of the 8, 7, 5 and 4 follow from the four digits case. The 2 could either make 842×75 or 84×752. In the second case 2 is multiplied by 84 and in the first by 75, so the second is better.

The smallest result is with 478×25.

B7 The largest result is $7 \times 63 \times 54 = 23\,814$.
The smallest is $3 \times 46 \times 57 = 7866$

B8 $415 \times 95 = 39\,425$

ℂ Division and remainders (p 29)

C1 25

C2 18

C3 29

C4 17

C5 19

C6 The pupil's problem

C7 15

𝔻 Division by a two-digit number (p 30)

D1 (a) 7 rem 6   (b) 20 rem 19
     (c) 6 rem 19   (d) 12 rem 12
     (e) 10 rem 13

D2 6

D3 14 crates, 11 bottles left over

D4 £1352

D5 The pupil's problems

D6 The pupil's problems

𝔼 Problems (p 31)

E1 421

E2 369

E3 11

E4 13

E5 570

E6 (a) Row 5, column 7
     (b) Row 16, column 1
     (c) Row 28, column 16
     (d) Row 50, column 7

Divide the number by 17. Adding 1 to the result gives the row number. The remainder gives the column number.

E7 (a) 63   (b) 160   (c) 387

Subtract 1 from the row number and multiply by 17. Then add the column number.

𝔽 Remainders with a calculator (p 32)

F1 10

F2 11

F3 44 boxes, 9 eggs left over

F4 (a) 29   (b) 46

Faulty calculator

There are various methods, involving trial and improvement, the idea of division as repeated multiplication and an understanding of multiplication by powers of 10.

What progress have you made? (p 33)

1 (a) 5   (b) 17

2 (a) 6 rem 3   (b) 8 rem 22

3 (a) 8   (b) 1

The mixed-up table puzzle (p 33)

Q 9, R 8, S 5, T 1, U 0, V 3, W 7, X 4, Y 6, Z 2

Practice booklet

Section B (p 9)

1 $24 \times 36 = 864$
  $58 \times 63 = 3654$
  $14 \times 46 = 644$
  $38 \times 14 = 532$
  $28 \times 24 = 672$
  $46 \times 36 = 1656$

2 (a) £10.92   (b) £15.48

3 (a) $34 \times 23 = \mathbf{782}$
  (b) $27 \times 3\mathbf{1} = \mathbf{8}37$
  (c) $72 \times 1\mathbf{9} = 1\mathbf{3}68$

4 (a) $72 \times 54 = 3888$
  (b) $25 \times 47 = 1175$
  (c) $74 \times 25 = 1850$

5 26 and 27

Section C (p 10)

1 7 packets

2 4 packets

3 4 boxes

4 17 plates

5 Each boy gets more (95p as against 85p).

6 (a) 38 apples   (b) 3 more apples

7 23 bucketfuls

8 19 times

Section D (p 11)

1 (a) 13 rem 1   (b) 12 rem 13
  (c) 11 rem 16   (d) 11 rem 12
  (e) 12 rem 39   (f) 17 rem 10
  (g) 22 rem 16   (h) 27 rem 9

2 8 coaches

3 (a) £108   (b) £23

4 (a) 20 crates   (b) 28 crates

5 19p

6 23p

Section E (p 12)

1 (a) 9 boxes   (b) 16 spare discs

2 28

3 35 bricks

4 Rows of 24 give the smallest number of empty seats (11).

5 214 rings

6 61 apples

7 13 grandchildren

Section F (p 13)

1 (a) 6   (b) 60   (c) 55
  (d) 73   (e) 81   (f) 1

2 The cheapest way is for the company to buy 28 boxes (£8400 for 2016 mice).
  To get exactly 2000 mice, buy 27 boxes (1944 mice) and 56 extra mice (£8408 for 2000 mice).

3 1041 postcards

4 (a) 33   (b) 1

5 (a) $a = 15$ (b) $b = 373$
  (c) $c = 810$, $d = 26$ or
    $c = 841$, $d = 27$ or
    $c = 872$, $d = 28$
  (d) $e = 90$ (e) $f = 168$
  (f) $g = 500$

④ Growing patterns

7S/5

Pupils investigate sequences arising from a variety of contexts.

The emphasis is on finding a rule to continue a sequence and explaining why the rule is valid. There is a little work on finding a rule that effectively gives the *n*th term (though not expressed in that way), but this aspect is covered more fully later.

Pupils should realise that just spotting a pattern in the first few numbers in a sequence (for example, 'add 3 to the previous number') is not enough to prove that the sequence will continue in the same way. You have to go back to the context (rose bushes, ponds, earrings, ...) to give a convincing explanation.

T	p 34 **A** Coming up roses	Investigation leading to a linear sequence
		Explaining why the sequence continues like this
	p 34 **B** Pond life	Investigation leading to a linear sequence
	p 35 **C** Changing shape	Consolidating work on linear sequences
T	p 36 **D** Earrings	Investigation leading to other kinds of sequence
	p 37 **E** Staircases	Investigation leading to a Fibonacci sequence

Optional Red and yellow multilink cubes Sheets 71 and 72
Practice booklet pages 14 and 15

A Coming up roses (p 34)

◊ A gardener is designing a display with red and white rose bushes planted as single rows of red roses, with a row of white roses on each side and a white rose at each end. Some examples are:

℡ represents a white rose bush.

℡ represents a red rose bush.

◊ To start with, pupils could think about the designs on the page and then draw a similar arrangement that uses a different number of red roses.

Encourage pupils to simplify the diagrams, for example by using coloured circles or the letters R and W.

Pupils count the number of red and white rose bushes needed in each arrangement so far and collect their results together in an ordered table. Discuss the advantages of tabulating in this way.

◊ Ask pupils to complete the table up to, say, 8 red rose bushes.

In discussion, bring out the fact that the number of white rose bushes increases by 2 for every extra red rose bush. Ask the pupils to use the diagrams to explain why this is so.

Emphasise that merely finding a pattern in the table is not explaining **why** the pattern is there and will continue.

◊ Now ask pupils to think about a larger number of red bushes, for example 20 red bushes.

If pupils use the rule that the number of white bushes is
(2 × the number of red bushes) + 2, ask them to explain why they know their rule works by referring to the arrangement of rose bushes. Emphasise that just because their rule works for a few results it does not follow it will work for all results.

◊ Pose a question like 'How many **red** bushes are needed for 158 **white** bushes?' Ask pupils how they work out the answer.

B Pond life (p 34)

C Changing shape (p 35)

As an extension, pupils could consider triangular ponds surrounded by triangular slabs.

D Earrings (p 36)

> Optional: Red and yellow multilink cubes

◊ Multilink has been found to be a very useful way to 'build' the earrings. It allows easy identification of duplicates and collection of results.

If pupils use multilink, make sure they realise that these two designs are different.

◊ Ask pupils to find as many different three-bead earrings as they can.

Collect the results for the whole class and discuss how they can be sure that they have found all possible designs for three beads.

The 8 different designs are:

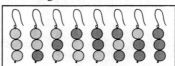

D4 Pupils cannot claim to be sure about the number of five-bead earrings until they have found them all (and shown no more exist) or until they have explained why the numbers in the sequence double each time.

D5 In (b), most pupils will find it difficult to explain why the number of earrings doubles each time. If so, encourage them to look at their sets of earrings for say 3 beads and 4 beads and to consider how they are related.

E Staircases (p 37)

Pupils investigate a situation that gives rise to a Fibonacci sequence.

> Optional: Sheets 71 and 72 (for recording results)

'Pupils gained a great deal from doing this on a real flight of stairs, as it was easy to see which moves were not allowed.'

◊ Emphasise that the staircase in the diagram has four steps and not five. This can confuse pupils and using sheets 71 and 72 may help.

◊ Ask pupils to think carefully about ways of recording their results. Pupils who choose to record their results as sequences of 1s and 2s may realise that the problem reduces to that of finding how many different ways a total can be reached by adding 1s and 2s.

E3 Cover up the last two entries in the table and ask pupils to imagine they were trying to predict the number of ways to climb four steps. It's very tempting to predict 4 ways, the sequence 1, 2, 3, … increasing by 1 each time. However, the prediction would be incorrect. This is an opportunity to reinforce the dangers of relying on an apparent number pattern.

B Pond life (p 34)

B1 The numbers of slabs in the table are 8, 12, 16, 20, 24, 28.

B2 44 slabs
Pupils' methods are likely to involve
• counting on in 4s or
• multiplying by 4 and adding 4 or
• adding 1 and multiplying by 4

B3 11 metres
Pupils' methods could involve
• extending the pattern in the table or
• working from the fact that a 10 by 10 pond needs 44 slabs or
• subtracting 4 and dividing by 4 or
• dividing by 4 and subtracting 1

B4 (a) The number of slabs needed goes up by 4 each time.

(b) The pupil's explanation: for example, an increase of 1 metre in width means an extra slab for each edge.
Since the pond has 4 edges, 4 extra slabs are needed.

B5 276

B6 B: $(n \times 4) + 4$

C Changing shape (p 35)

C1 (a) (i) 20 slabs   (ii) 14 slabs
(b) The numbers of slabs in the table are 12, 14, 16, 18, 20, 22, 24.

C2 (a) The number of slabs needed goes up by 2 each time.

(b) The pupil's explanation

C3 31 metres

C4 C: $(x \times 2) + 10$

D Earrings (p 36)

D1 4 different earrings

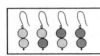

D2 The numbers of different earrings in the table are 2, 4, 8.

D3 (a) 16 different earrings

(b) The pupil's method: for example, $8 \times 2 = 16$ or $8 + 8 = 16$

(c) One way to organise the results is to add a yellow bead to each of the three-bead earrings and then add a red bead.

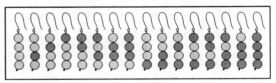

Another way is to look at earrings with 0 red beads, 1 red bead, 2 red beads, …

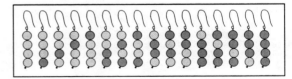

D4 32 different earrings

D5 (a) The number of earrings doubles each time.

(b) The pupil's explanation

D6 6

D7 24

D8 (a) The number of earrings in the table are 1, 2, 6, 24, 120

(b) Multiply by 2, then 3, then 4, then 5, …

E Staircases (p 37)

E1 5 ways

E2 (a) 8 ways

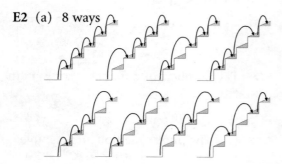

(b) 3 ways

E3 The numbers of different ways in the table are 1, 2, 3, 5, 8.

E4 (a) 13 ways

(b) The pupil's method: for example, add the previous two numbers in the sequence, 5 + 8 = 13.

(c)

E5 To find the number of ways to climb a staircase, add the numbers of ways to climb the previous two staircases.

E6 The pupil's explanations. A possible explanation is as follows: Consider for example a staircase with 5 steps. Each set of 'steps' up the staircase ends in a '1-step' or a '2-step'. Find the ways that end in a '1-step' by considering all the ways to climb 4 steps (add a '1-step' on to each of them). Find the ways that end in a '2-step' by considering all the ways to climb 3 steps (add a '2-step' on to each of them). So find the total number of ways to climb 5 steps by adding the number of ways to climb 4 steps to the number of ways to climb 3 steps. A similar argument works for any size staircase.

*E7

Number of steps	1	2	3	4	5	6
Number of ways	1	2	4	7	13	24

Each number of ways is found by adding the previous three.

What progress have you made? (p 38)

1 The numbers of yellow rose bushes in the table are 5, 6, 7, 8, 9.

2 (a) The number of yellow bushes goes up by 1 each time.

(b) The pupil's explanation: for example, an increase of 1 red bush means an extra yellow bush in the bottom row.

3 (a) 104

(b) The pupil's method: for example, 100 + 4 = 104

4 (a) 46

(b) The pupil's method: for example, 50 − 4 = 46

5 C: $n + 4$

6 B: $(n \times 2) + 8$

Practice booklet

Sections B and C (p 14)

1. (a) The pupil's drawing; 16 corners
   (b) The numbers of corners in the table are 4, 7, 10, 13, 16, 19.
   (c) The number of corners goes up by 3 for each added square.
   (d) One corner of a new square that joins the chain has already been counted, so 3 corners are added each time.
   (e) 31 corners
   (f) C: $(n \times 3) + 1$

2. (a) The pupil's sketch; 14 corners
   (b) The numbers of corners in the table are 8, 11, 14, 17, 20.
   (c) B: $(n \times 3) + 2$

Sections D and E (p 15)

1. (a) 30
   (b) The numbers of squares in the table are 1, 5, 14, 30, 55.
   (c) The next square number is added each time.
   For example, from the 3 by 3 grid to the 4 by 4 grid the number of squares goes up by 4 by 4, or 16.
   (d) The number of squares in a 3 by 3 grid, for example, is $1 + 4 + 9$, or $1^2 + 2^2 + 3^2$.
   For a 4 by 4 grid the number will be $1^2 + 2^2 + 3^2 + 4^2$, an increase of 4^2.

2. (a) 8
   (b) The numbers of arrangements in the table are 1, 2, 4, 8, 16.
   (c) The number of arrangements doubles each time.
   (d) The pupil's explanation

Test it!

Pupils collect measurements to test general statements.
Higher attainers go on to find an approximate relationship between
two sets of measurements.

T

p 39	**A** I don't believe it!	Planning a task Collecting data to test a statement Measuring
p 40	**B** Organising your results	Organising measurements
p 42	**C** Now it's your turn!	Testing a chosen statement
p 42	**D** Number detective	Using division to find an approximate connection between two sets of measurements

---

Essential

Metre sticks, tape measures (enough for at least one item per group)

🅐 I don't believe it! (p 39)

Metre sticks, tape measures (at least one item per group)

T

◊ Organise pupils into small working groups. (Groups of four work well.)
After you have introduced the statement 'Everyone is six and a half feet
tall', the groups can discuss the first set of questions.

It should become clear that

- six and a half feet tall means six and a half 'foot lengths' tall

- the statement can be tested by measuring or simply stepping off each
pupil's foot length against their height

You could have a general discussion at this point comparing the groups'
plans for testing the statement. Or the groups could move straight into
carrying out their plans. Some pupils may need a lot of help with
measuring.

Recording of data may be haphazard. This is taken up in section B, which
some teachers have preferred to do before A.

Methods used by pupils include:

- making an outline of themselves on a roll of old wallpaper, then cutting out their footprints to test the statement
- Blu-tacking rulers to walls to make it easier to measure heights
- measuring out heights on the tape and then 'stepping off'

Six and a half feet tall

'Pupils enjoyed this topic. However, there were several teething problems such as which groups they were in and who was responsible for what. When things settled down they produced some excellent work which was good for display. The groups also gave a presentation of their work.'

In work of this kind it is important to make a plan and to adapt it as necessary. In many cases pupils do not see the need for a plan, preferring instead to 'jump right in'. You may find examples to emphasise this point in the pupils' own work.

Encourage each group to compare findings with others.

◊ After this discussion it is worth raising the issue of whether the statement 'Everyone is six and a half feet tall' is true for a wider population. Pupils could investigate the statement by measuring younger or older people at home for homework.

B Organising your results (p 40)

◊ Some teachers have preferred to do this section before section A.

B1 Pupils should consider the problems of
- mixed units
- writing results in different orders

B2 You may need to help with 'approximately'.

C Now it's your turn! (p 42)

Pupils choose their own general statement to test.

◊ Each group must decide how to measure, for example, the 'length' of a person's head.

◊ A formal write-up is not necessarily expected at this stage. You could ask for a poster from each group or each pupil. (Later in the course there is a more specific focus on writing up results.)

Ⓓ **Number detective** (p 42)

Pupils try to find an approximate relationship between two sets of measurements.

◊ This is suitable for pupils working in pairs.

D2 Some pupils may discover that 1.9 is a slightly better multiplier than 2.

D3 If pupils suggest other relationships, by all means let them investigate.

Ⓑ **Organising your results** (p 40)

B1 This method can be improved by presenting results in the same order and everyone using the same units.

B2 Ben is correct.

B3 (a) It may mean the height from the chin to the top of the head (when the mouth is closed).

(b) Tim's arm span might be 170 cm.
Gina's height might be 1.44 m.
Sue's height might be 1.59 m.
Sue's foot length might be 26 cm.
Ryan's height might be 1.5 m.
Lara's hand span might be 17 cm.

(c) Tim, Sue, Ryan (if he is 1.5 m tall) and Majid can go on the rides.

(d) Neena's arm span might be about the same length as Ajaz's, 153 cm.

(e) Depends on the pupil's own measurements

(f) About 6.4 to 7.3 times
(The height and head length have to be expressed in the same units before dividing.)

Ⓓ **Number detective** (p 42)

D1 (a) 4

(b)

| | Elbow | |
Height (cm)	distance (cm)	Height ÷ arm
160	40	4
136.5	35	3.9
175.5	45	3.9
161	41	3.92...
133	35	3.8
152	42	3.61...

(c) The height is roughly four times the distance from elbow to finger tip.

D2 The approximate rule is that the height is twice the distance round foot.

D3 The pupil's own work

What progress have you made? (p 43)

1 The height is about 9 times the hand length.

Number patterns

p 44   **A** Exploring a number grid

p 45   **B** Square numbers

p 46   **C** Cubes

p 47   **D** Square roots

p 49   **E** Sequences

p 50   **F** Sequences in tables

p 51   **G** Even number, odd number

Optional

Sheet 96
Multilink or other cubes

Practice booklet pages 16 and 17

A **Exploring a number grid** (p 44)

These investigations are all based on the six-column grid and are graded in difficulty. Some teachers have preferred to do some or all of them later in the unit.

Optional: Sheet 96

Investigation 1 (Add a number in column A to one in column B)

◊ This is a good one to start on with the whole class. It leads on to variations which pupils can investigate for themselves.

The result of A + B is always in column C (provided the grid is extended downwards). You can then ask pupils what they think could be meant by 'investigate further'. All suggestions should be responded to positively, even though they may lead nowhere.

Here are some fruitful suggestions:

What if we add two different columns?
What if we subtract?
What if we multiply?

Investigation 2 (Multiples)

◊ These occur in spatially regular patterns.

Investigation 3 (Predict the 30th number in column B, etc.)

◊ There are various ways to do this. One is to work out the 30th number in column F ($30 \times 6 = 180$) and then work back to 176.

Investigation 4 (Predict which column 500 will be in, etc.)

◊ $500 \div 6 = 83$ remainder 2, so 500 will be in column 2.

Investigation 5 (Prime numbers)

◊ Prime numbers (except for 2 and 3) are only in columns A and E. This is because the numbers in the other columns are either even or multiples of 3.

B Square numbers (p 45)

C Cubes (p 46)

Optional: Multilink or other types of cube

D Square roots (p 47)

E Sequences (p 49)

F Sequences in tables (p 50)

G Even number, odd number (p 51)

B Square numbers (p 45)

B1 $6 \times 6 = 36$ and $7 \times 7 = 49$

B2 64, 81, 100

B3 3, 5, 7, 9, … odd numbers

B4 (a) 16   (b) 25   (c) 121 (d) 400

B5 Rob has done 10×2. He should have done 10×10.

B6 (a) 13   (b) 33   (c) 73
      (d) 61   (e) 155

B7 $6 = 2^2 + 1^2 + 1^2$

$7 = 2^2 + 1^2 + 1^2 + 1^2$

$8 = 2^2 + 2^2$

$9 = 3^2$

$10 = 3^2 + 1^2$

$11 = 3^2 + 1^2 + 1^2$

$12 = 2^2 + 2^2 + 2^2$

$13 = 3^2 + 2^2$

$14 = 3^2 + 2^2 + 1^2$

$15 = 3^2 + 2^2 + 1^2 + 1^2$

$16 = 4^2$

$17 = 4^2 + 1^2$

$18 = 3^2 + 3^2$

$19 = 3^2 + 3^2 + 1^2$

$20 = 4^2 + 2^2$

$21 = 4^2 + 2^2 + 1^2$

$22 = 3^2 + 3^2 + 2^2$

$23 = 3^2 + 3^2 + 2^2 + 1^2$

$24 = 4^2 + 2^2 + 2^2$

$25 = 5^2$

$26 = 5^2 + 1^2$

$27 = 5^2 + 1^2 + 1^2 \ \text{or} \ 3^2 + 3^2 + 3^2$

$28 = 5^2 + 1^2 + 1^2 + 1^2 \ \text{or}$

$\qquad 4^2 + 2^2 + 2^2 + 2^2 \ \text{or}$

$\qquad 3^2 + 3^2 + 3^2 + 1^2$

$29 = 5^2 + 2^2$

$30 = 5^2 + 2^2 + 1^2$

B8 (a) $7^2 = 1 + 3 + 5 + 7 + 9 + 11 + 13$

(b) 400

(c) $9^2 = 1 + 2 + 3 + 4 + 5 + 6 + 7 + 8 +$
$\qquad 9 + 8 + 7 + 6 + 5 + 4 + 3 + 2 + 1$
$\qquad 9^2 = 1 + 8 + 16 + 24 + 32$

ℂ Cubes (p 46)

C1 (a) 27   (b) 3^3

C2 (a) 72   (b) 150   (c) 152   (d) 657

C3 The differences are 7, 19, 37, 61, 91, …
The differences between the differences
are 12, 18, 24, 30, …
These are the multiples of 6, starting
with 12.

C4 3375 (15^3)

𝔻 Square roots (p 47)

D1 (a) 3   (b) 5   (c) 10

(d) 8   (e) 1

D2 (a) 2   (b) 9   (c) 6

D3 (a) 7   (b) 12   (c) 21

(d) 123

D4 31

D5 (a) 3000

(b) The answer depends on how many
people can occupy a square metre.
If it is 1, then the length of the side
would be 3 km.

D6 On the same basis, the area would be
6400 km^2 (side 80 km).

Trial and improvement working should be
shown in questions D7–D9.

D7 $37^2 = 1369$, so $\sqrt{1369} = 37$

D8 (a) 27   (b) 41   (c) 57   (d) 921

D9 $\sqrt{20} = 4.472\,135\,95\ldots$

D10 (a) 300 km

(b) 629 m = 0.629 km, so you can see
79.3 km

True or false?

1 False; for example, $\sqrt{0.25} = 0.5$

2 False; for example, $0.5^2 = 0.25$

3 False; $2.5^2 = 6.25$, which is not halfway
between 4 and 9

E Sequences (p 49)

E1 (a) 25   (b) Add 4

E2 (a) Subtract 3   (b) Multiply by 3

E3 (a) 7, **11**, 15, **19**, 23, 27, **31**   add 4

    (b) 1, **4**, 7, **10**, **13**, 16, **19**   add 3

    (c) **41**, **37**, 33, **29**, 25, **21**, 17   subtract 4

    (d) 44, **39**, 34, **29**, **24**, 19, **14**   subtract 5

E4 (a) 21, 28, 36

    (b) Add 2, then 3, then 4, …
      (one more each time)

E5 You get the sequence of square numbers.

For example, the fourth and fifth
triangle numbers give 5^2:

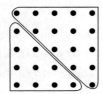

E6 Add 4, then 7, then 10, …
    (three more each time)

F Sequences in tables (p 50)

F1 (a) Square numbers

    (b) Square numbers + 1

    (c) Add 2, then 4, then 6, then 8, …
      (add even numbers)

F2 (a) Add 4, then 8, then 12, …
      (add multiples of 4);
      ?, **?** are 61, 85.

    (b) The pupil's investigation

F3 On every diagonal the differences go up
by 8 each time.

    81, 121 (odd square numbers)
    65, 101 (even squares + 1)
    57, 91
    73, 111

G Even number, odd number (p 51)

G1 An odd number consists of a set of 2s
and an extra 1.
When two odd numbers are added, the
extra 1s pair off and so the result is even.

For example:

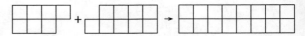

G2 Two consecutive numbers are either odd,
even or even, odd. In both cases the sum
will be odd.

G3 (a) Odd

    (b) The pattern is: odd, odd, even, even,
      odd, odd, even, even, …

    (c) Consecutive numbers are alternately
      odd, even.
      When we add up, we get

      1   odd
      1 + 2   odd + even = odd
      1 + 2 + 3   odd + odd = even
      1 + 2 + 3 + 4   even + even = even
      and so on

What progress have you made? (p 51)

1 (a) 121   (b) $12^2 = 144$

2 53.09…

3 64

4 66

5 The pupil's investigation

Practice booklet

Sections B, C and D (p 16)

1 (a) 9   (b) 49   (c) 81   (d) 10 000

2 (a) 27   (b) 56   (c) 257   (d) 19

3 (a) 27   (b) 1000

    (c) 1331   (d) 125

4 44 $(6^2 + 2^3)$, 64 $(10^2 - 6^2)$, 125 (5^3),
128 $(4^3 + 4^3)$, 225 (15^2), 243 $(3^3 + 6^3)$,
271 $(10^3 - 9^3)$

5 (a) 1936 (44×44)

    (b) 1728 $(12 \times 12 \times 12)$

6 (a) 4   (b) 7   (c) 2

    (d) 12   (e) 20

7 4000

8 (a) 2 seconds

    (b) About 6 seconds (6.32…)

Sections E and F (p 17)

1 (a) Add 11   (b) Subtract 6

    (c) Multiply by 3   (d) Divide by 2

2 (a) 100   (b) 4

    (c) 1215   (d) 15

3 (a) 3, 9, 18

    (b) The pupil's sketch; 30

    (c) Each number in the sequence is 3 times the corresponding triangle number.

    (d) The reason is shown most easily by splitting up the diagrams.

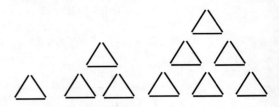

4 (a) 66   (b) 207

    (c) (i) 37th row, 2nd column

        (ii) 62nd row, 3rd column

Angles and triangles

p 52	**A** Drawing a triangle accurately	Constructing a triangle given three sides, using compasses
p 54	**B** Equilateral triangles	Making a net for an octahedron
p 56	**C** Isosceles triangles	
p 58	**D** Scalene triangles	
p 59	**E** Using angles	Constructing a triangle given sides and angles
p 62	**F** Calculating angles	Angles round a point and on a straight line
p 64	**G** Angles of a triangle	
p 66	**H** Using angles in isosceles triangles	

Essential

Plain paper
Compasses
Scissors
Tracing paper
An envelope to keep cut or traced triangles
Thin card (for nets)
Glue
Angle measurer

Practice booklet pages 18 to 21

Optional

Compasses for board or OHP
Triangular dotty paper (sheet 2)

🅐 Drawing a triangle accurately (p 52)

Plain paper, compasses, tracing paper or scissors

◊ A good way to begin is by sketching an 8, 5, 10 triangle on the board and challenging the class to draw it accurately with pencil and ruler only. They may get an accurate result, but probably only after some trial and error. This should help them see the advantage of using compasses.

◊ If there are discrepancies when pupils compare their triangles with their neighbours', the problems should be sorted out and the triangles drawn again. They will be needed for later questions.

Investigation

For three lengths to make a triangle, the longest must be less than the sum of the other two.

Ⓑ **Equilateral triangles** (p 54)

> The triangles made in section A, compasses, scissors, thin card, glue, (possibly) triangular dotty paper

◊ Pupils who, because of poor manipulative skills, are likely to be discouraged rather than helped by doing so much work with compasses could work on triangular dotty paper.

◊ Pupils could design and make other polyhedra with equilateral triangles as faces. Here is an easy-to-make net for a regular icosahedron.

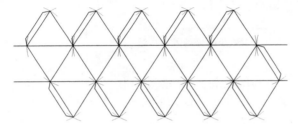

Ⓒ **Isosceles triangles** (p 56)

> The triangles made in section A, compasses, scissors, glue

◊ The fact that an equilateral triangle is a special case of an isosceles triangle may come up in the answers to questions and in discussion. There is no need to make a big thing of it at this stage.

C5 You may need to give help on naming triangles by the letters of their vertices.

Ⓓ **Scalene triangles** (p 58)

> The triangles made in section A

◊ Ensure pupils realise that a right-angled triangle can be scalene.

E Using angles (p 59)

This section includes drawing a triangle given one side and two angles, and given two sides and the angle between them.

Angle measurer

F Calculating angles (p 62)

These calculations use angles round a point, angles on a line and vertically opposite angles.

High attainers can make up their own questions similar to F5 and give them to others to do.

G Angles of a triangle (p 64)

Angle measurer, scissors

◊ Ask everybody to draw a triangle, measure its angles and add them together. Enough of the results should be close to 180° to blame the discrepancy on inaccurate drawing and measurement!

◊ For the torn-off angles demonstration you may need to remind pupils about angles on a line.

Of course, neither of these approaches amounts to a proof. The proof is taken up in later material.

H Using angles in isosceles triangles (p 66)

H8 Part (b) can be discussed with pupils when you draw together the work in the unit.

B Equilateral triangles (p 54)

B1 (a) Triangles C and F

(b) They are all 60°.

B2 Yes, you can fold it so two sides and two angles match up with one another. The fold line is the mirror line.

B3 6

B4 12

B5 All the angles and sides of the triangular 'faces' are the same.

C Isosceles triangles (p 56)

C1 Triangles B, C, E and F
(C and F are also equilateral).

C2 (a) It is 90° (a right angle).

(b) They are equal.

(c) It has reflection symmetry, with the fold line as a line of symmetry.

C3 The pupil's model

C4 Between 5 and 7 cm is a reasonable result (set squares are useful measuring aids). The important thing is that pupils don't just measure the sloping edge length or even the distance from the midpoint of the side of the square to the top of the vertex (a result of about 7.5 cm would indicate this).

C5 These triangles are isosceles. Finding just some of them is a fair achievement.

ACO, ECN, ACN, NCO, OCE, BOC, DNC, ANO, EON, FON, GNO, KGO, MFN, NIO, NLO, BCF, CDG, CFO, CGN, FIN, GIO

Diagonal cut puzzle

The triangles cut out are two pairs of isosceles triangles.

These are the two ways of fitting them together.

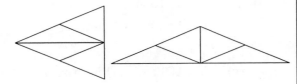

D Scalene triangles (p 58)

D1 Triangles A and D

D2 ABC, ABD, ACE, ADE

D3 (a) Isosceles   (b) Scalene
     (c) Equilateral   (d) Isosceles
     (e) Scalene   (f) Isosceles
     (g) Equilateral

D4 No. Pupils can think about what would happen if they tried folding so that one side went on to another. They would be different lengths so they would not match.

E Using angles (p 59)

E1 The pupil's triangle

E2 The pupil's triangles

E3 It is not possible to complete the triangle because two of the lines are parallel.

E4 The pupil's triangle

E5 The pupil's triangles

E6 (a) The pupil's triangle
     (b) BC = 7.5 cm, angle at B = 49°, angle at C = 61°

E7 (a) The pupil's triangle
     (b) XZ = 7.4 cm, YZ = 12.4 cm, angle at Z = 20°

E8 (a) The pupil's triangle
     (b) AC = 12.5 cm, angle at A = 61°, angle at C = 34°

E9 (a) The pupil's triangle
     (b) QR = 12.5 cm, angle at Q = 32°, angle at R = 23°

E10 It is impossible to draw a triangle (because the circle centre B with radius 5 cm does not cross the line through A at 40° to AB).

E11 Two different triangles can be drawn (because the circle centre Q with radius 7 cm crosses the line through P at two points).

F Calculating angles (p 62)

F1 $a = 80°$, acute $b = 145°$, obtuse
     $c = 60°$, acute $d = 250°$, reflex

F2 $a = 140°$ $b = 35°$ $c = 130°$ $d = 115°$

F3 $a = 39°$ $b = 43°$ $c = 97°$ $d = 234°$
     $e = 151°$

F4 $a = 110°$ $b = 70°$ $c = 110°$ $d = 17°$
     $e = 163°$ $f = 17°$ $g = 62°$ $h = 25°$
     $i = 93°$ $j = 25°$

F5 $a = 102°$ $b = 36°$ $c = 123°$ $d = 36°$
$e = 88°$ $f = 129°$ $g = 113°$ $h = 134°$
$i = 22°$ $j = 125°$ $k = 125°$

Ⓖ Angles of a triangle (p 64)

G1 (a) 50° (b) 60° (c) 55°
(d) 80° (e) 20°

G2 (a) 75° (b) 121° (c) 85°
(d) 66° (e) 91°

G3 Each angle is 60°, because 180 ÷ 3 = 60.

G4 (a) 60° (b) 25° (c) 17°
(d) 62° (e) 33°

G5 $a = 60°$ $b = 120°$ $c = 70°$
$d = 110°$ $e = 80°$

G6 $a = 70°$ $b = 65°$ $c = 128°$
$d = 80°$ $e = 100°$ $f = 148°$

G7 To draw the triangle, the third angle (40°) must be worked out first.

Ⓗ Using angles in isosceles triangles (p 66)

H1 They are equal.

H2 $a = 75°$ $b = 30°$ $c = 24°$ $d = 132°$
$e = 54°$ $f = 72°$ $g = 18°$ $h = 144°$
$i = 77°$ $j = 26°$

H3 $a = 70°$ $b = 70°$ $c = 40°$ $d = 40°$
$e = 66°$ $f = 66°$ $g = 30°$ $h = 30°$
$i = 48°$ $j = 48°$

H4 (a) Either 40° and 100° or 70° and 70°
(b) Either 72° and 36° or 54° and 54°
(c) 25° and 25°

H5 (a) 360° (b) 72°
(c), (d) The pupil's pentagon

H6 The pupil's drawing of a regular polygon

H7 The triangles will not fit together because 50° does not go into 360° exactly.

H8 (a) 9 sides
(b) The 'non-base' angle of the isosceles triangle must be a factor of 360°

H9 $a = 30°$ $b = 75°$ $c = 150°$

H10 $p = 108°$ $q = 144°$

*__H11__ (a) 108°, 36°, 36°
(b) 36°, 72°, 72°
(c) (i) The sides in contact are the same length, because the pentagon is regular.
(ii) The angles marked •,○ add up to 180°

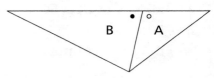

*__H12__ 24

What progress have you made? (p 69)

1 The pupil's triangles

2 (a) ABD
(b) EBD, ABD
(c) ABE, ADE, BEC, DEC, ABC, ADC
(d) ABC, EBC, ADC, EDC

3 $a = 47°$ $b = 110°$ $c = 70°$ $d = 101°$

4 $a = 112°$ $b = 57°$

5 $a = 67°$ $b = 46°$ $c = 38°$

Practice booklet

Sections A, B, C and D (p 18)

1 The pupil's drawing; 90°, 37°, 53°

2 The 8 cm side is longer than the total of the other two sides.

3 (a) ABE, ADE, BCE (b) CDE
(c) None (d) ABE

Section E (p 18)

1 (a) The pupil's accurate drawing
   (b) Angle C = 70°
       AC = 5.4 cm, BC = 3.9 cm

2 The remaining angles and sides are
   (a) Angle B = 38°, angle C = 30°,
       BC = 12.1 cm
   (b) Angle Z = 80°, XZ = 5.0 cm,
       YZ = 6.3 cm
   (c) Angle P = 37°, angle R = 53°,
       QR = 4.5 cm

Section F (p 19)

1 $a = 170°$ $b = 60°$ $c = 45°$
   $d = 36°$ $e = 130°$ $f = 105°$
   $g = 64°$ $h = 64°$ $i = 116°$
   $j = 60°$ $k = 25°$ $l = 65°$
   $m = 34°$ $n = 80°$ $o = 140°$
   $p = 96°$ $q = 75°$ $r = 72°$
   $s = 75°$ $t = 33°$

Section G (p 20)

1 $a = 60°$ $b = 40°$ $c = 25°$
   $d = 58°$ $e = 139°$ $f = 65°$
   $g = 115°$ $h = 49°$ $i = 131°$
   $j = 38°$ $k = 67°$ $l = 43°$

Section H (p 21)

1 $a = 74°$ $b = 74°$ $c = 55°$
   $d = 70°$ $e = 30°$

2 (a) 25°, 25°
   (b) Either 50° and 80° or 65° and 65°

3 $a = 45°$ $b = 67\frac{1}{2}°$

4 $x = 32°$ $y = 68°$

5 $a = 33°$

 **Fractions**

This unit revises equivalent fractions and simplifying and goes on to deal with ordering, adding and subtracting fractions and multiplying a fraction by a whole number. Examples involving mixed numbers are included.

> **Practice booklet** pages 22 and 23

Ⓐ Fractions of shapes (p 70)

Ⓑ Simplifying fractions (p 72)

> The questions include examples of changing decimals to fractions.

Ⓒ Comparing fractions (p 74)

Ⓓ Mixed numbers (p 74)

Ⓔ Adding and subtracting fractions (p 75)

> The initial approach here is visual, to bring out the need for common denominators.

Ⓕ Multiplying a fraction by a whole number (p 77)

> You could bring out the fact that, for example, '$\frac{3}{4} \times 8$' (thought of as 8 lots of $\frac{3}{4}$) and '$\frac{3}{4}$ of 8' lead to the same result.

Ⓖ Mixed questions (p 77)

Ⓐ Fractions of shapes (p 70)

A1 (a) $\frac{1}{8}$ (b) $\frac{3}{8}$ (c) $\frac{7}{8}$ (d) $\frac{5}{8}$

A2 (a) $\frac{1}{4}$ (b) $\frac{5}{16}$ (c) $\frac{6}{16}$ or $\frac{3}{8}$

A3 (a) $\frac{7}{16}$ (b) $\frac{3}{16}$ (c) $\frac{6}{9}$ or $\frac{2}{3}$

     (d) $\frac{3}{8}$ (e) $\frac{3}{16}$ (f) $\frac{1}{16}$

A4 Thailand $\frac{2}{6}$ or $\frac{1}{3}$ Czech Republic $\frac{3}{8}$

A5 (a) $\frac{7}{9}$ (b) $\frac{1}{2}$ (c) $\frac{5}{9}$ (d) $\frac{3}{24}$ or $\frac{1}{8}$

*A6 (a) $\frac{3}{4}$ of $\frac{4}{5}$

     (b) (i) $\frac{3}{4}$ of $\frac{1}{3}$ or $\frac{1}{3}$ of $\frac{3}{4}$

        (ii) $\frac{4}{5}$ of $\frac{2}{3}$ or $\frac{2}{3}$ of $\frac{4}{5}$

     (c) (i) $\frac{5}{7}$ (ii) $\frac{1}{4}$

Ⓑ Simplifying fractions (p 72)

B1 (a) $\frac{1}{2}$ (b) $\frac{3}{4}$ (c) $\frac{1}{4}$
     (d) $\frac{3}{4}$ (e) $\frac{3}{4}$

B2 (a) $\frac{1}{4}$ (b) $\frac{3}{8}$ (c) $\frac{2}{3}$
     (d) $\frac{3}{7}$ (e) $\frac{1}{5}$

B3 (a) $\frac{2}{3}$ (b) $\frac{5}{8}$ (c) $\frac{2}{5}$
     (d) cannot be simplified   (e) $\frac{2}{5}$

B4 (a) $\frac{1}{3}$ (b) –   (c) $\frac{2}{3}$ (d) –
     (e) $\frac{2}{5}$ (f) $\frac{2}{3}$ (g) $\frac{3}{7}$ (h) –
     (i) $\frac{5}{8}$ (j) $\frac{4}{15}$

B5 $\frac{2}{3}$

B6 (a) $\frac{1}{12}$ (b) $\frac{1}{4}$ (c) $\frac{1}{3}$ (d) $\frac{7}{16}$
     (e) $\frac{5}{8}$ (f) $\frac{35}{48}$ (g) $\frac{5}{6}$

B7 $\frac{3}{20}$

B8 (a) $\frac{6}{25}$ (b) $\frac{3}{50}$ (c) $\frac{19}{20}$ (d) $\frac{33}{100}$
     (e) $\frac{8}{25}$ (f) $\frac{4}{5}$ (g) $\frac{9}{20}$ (h) $\frac{3}{25}$
     (i) $\frac{17}{20}$ (j) $\frac{1}{50}$

B9 $\frac{33}{200}$

B10 (a) $\frac{11}{40}$ (b) $\frac{1}{8}$ (c) $\frac{37}{250}$
     (d) $\frac{11}{200}$ (e) $\frac{1}{125}$

B11 (a) fiftieths   (b) twenty-fifths
     (c) fortieths

*B12 $\frac{9}{20}$

Ⓒ Ordering fractions (p 74)

C1 (a) $\frac{3}{8}$ (b) $\frac{2}{3}$ (c) $\frac{4}{5}$ (d) $\frac{5}{8}$
     (e) $\frac{4}{9}$ (f) $\frac{3}{8}$ (g) $\frac{3}{5}$ (h) $\frac{9}{20}$

Ⓓ Mixed numbers (p 74)

D1 (a) $\frac{4}{3}$ (b) $\frac{9}{4}$ (c) $\frac{17}{5}$
     (d) $\frac{17}{10}$ (e) $\frac{23}{8}$

D2 (a) $2\frac{1}{2}$ (b) $1\frac{2}{3}$ (c) $1\frac{3}{4}$
     (d) $2\frac{1}{5}$ (e) $2\frac{3}{5}$

Ⓔ Adding and subtracting fractions (p 75)

E1 (a) $\frac{7}{12}$
     (b) Because 12 is divisible by both 3 and 4 (and 13 isn't)

E2 (a) $\frac{5}{6}$ (b) $\frac{11}{12}$ (c) $\frac{1}{2}$ or $\frac{3}{6}$
     (d) $\frac{5}{12}$ (e) $\frac{1}{3}$ or $\frac{4}{12}$

E3 The pupil's (optional) strip, $\frac{7}{10}$

E4 (a) $\frac{9}{10}$ (b) $\frac{1}{2}$ or $\frac{5}{10}$ (c) $\frac{9}{20}$
     (d) $\frac{19}{20}$ (e) $\frac{19}{24}$

E5 (a) $\frac{8}{15}$ (b) $\frac{11}{15}$ (c) $\frac{7}{8}$
     (d) $\frac{29}{30}$ (e) $\frac{47}{60}$

E6 (a) $\frac{1}{6}$ or $\frac{2}{12}$ (b) $\frac{5}{12}$ (c) $\frac{1}{3}$ or $\frac{2}{6}$ or $\frac{4}{12}$
     (d) $\frac{3}{12}$ or $\frac{1}{4}$ (e) $\frac{2}{12}$ or $\frac{1}{6}$

E7 The pupil's (optional) strip, $\frac{5}{24}$

E8 (a) $\frac{2}{15}$ (b) $\frac{1}{15}$ (c) $\frac{3}{16}$
     (d) $\frac{9}{20}$ (e) $\frac{7}{10}$

E9 (a) $\frac{19}{30}$ (b) $\frac{1}{30}$ (c) $\frac{17}{40}$
     (d) $\frac{1}{24}$ (e) $\frac{13}{60}$

E10 (a) $\frac{19}{24}$ (b) $\frac{5}{8}$ (c) $\frac{31}{40}$
     (d) $\frac{19}{24}$ (e) $1\frac{1}{20}$

E11 (a) $2\frac{3}{8}$ (b) $4\frac{5}{12}$ (c) $2\frac{1}{12}$
     (d) $2\frac{7}{24}$ (e) $4\frac{1}{12}$

E12 (a) $\frac{13}{24}$ (b) $\frac{19}{40}$ (c) $\frac{11}{20}$
     (d) $\frac{11}{24}$ (e) $\frac{7}{20}$

E13 (a) $\frac{5}{8}$ (b) $\frac{17}{20}$ (c) $1\frac{1}{2}$
     (d) $\frac{7}{12}$ (e) $1\frac{1}{24}$

Printer's fractions

It is helpful to think of a number which is divisible by 2, 3, 4 and 5. The smallest such number is 60. So think of an em as 60 parts. Then en = 30, thick = 20, mid = 15, thin = 12.

These are all the possible spaces that could be made:

12 thin

15 mid

20 thick

24 2 thins

27 mid + thin

30 en, or 2 mids

32 thick + thin

35 thick + mid

36 3 thins

39 mid + 2 thins

40 2 thicks

42 en + thin, or 2 mids + thin

44 thick + 2 thins

45 en + mid, or 3 mids

47 thick + mid + thin

48 4 thins

50 en + thick, or 2 mids + thick

51 mid + 3 thins

52 2 thicks + thin

54 en + 2 thins, or 2 mids + 2 thins

55 2 thicks + mid

56 thick + 3 thins

57 en + mid + thin, or 3 mids + thin

59 thick + mid + 2 thins

60 em, or 2 ens, or 3 thicks, or 4 mids,
   or en + 2 mids or 5 thins

𝔽 Multiplying a fraction by a whole number (p 77)

F1 (a) $\frac{12}{5}$ or $2\frac{2}{5}$ (b) $\frac{32}{5}$ or $6\frac{2}{5}$ (c) $\frac{30}{8}$ or $3\frac{3}{4}$
   (d) $\frac{35}{6}$ or $5\frac{5}{6}$ (e) $\frac{25}{8}$ or $3\frac{1}{8}$

F2 (a) $1\frac{1}{2}$ (b) $4\frac{4}{5}$ (c) $7\frac{1}{2}$
   (d) $3\frac{3}{4}$ (e) $1\frac{1}{4}$

𝔾 Mixed questions (p 77)

G1 (a) $\frac{1}{20}$ (b) $1\frac{5}{24}$ (c) $1\frac{7}{8}$ (d) $\frac{9}{20}$

G2 (a) $10\frac{1}{2}$ (b) $1\frac{19}{30}$ (c) $1\frac{11}{12}$ (d) $2\frac{7}{20}$

What progress have you made? (p 77)

1 (a) $\frac{4}{5}$ (b) $\frac{3}{4}$

2 $\frac{5}{8}$ ($\frac{5}{8} = \frac{35}{56}$, $\frac{4}{7} = \frac{32}{56}$)

3 (a) $\frac{13}{5}$ (b) $9\frac{1}{3}$

4 (a) $\frac{29}{40}$ (b) $\frac{5}{12}$ (c) $1\frac{7}{12}$

5 (a) $5\frac{1}{4}$ (b) $7\frac{1}{5}$

Practice booklet

Sections A and B (p 22)

1 (a) $\frac{6}{13}$ (b) $\frac{7}{16}$ (c) $\frac{1}{6}$

2 (a) $\frac{1}{4} = \frac{\mathbf{8}}{32}$ (b) $\frac{2}{3} = \frac{\mathbf{14}}{21}$ (c) $\frac{3}{5} = \frac{\mathbf{27}}{45}$

3 (a) $\frac{3}{7}$ (b) $\frac{5}{8}$ (c) $\frac{5}{8}$
   (d) $\frac{1}{5}$ (e) $\frac{1}{3}$

4 (a) $\frac{11}{25}$ (b) $\frac{17}{50}$ (c) $\frac{1}{25}$
   (d) $\frac{11}{20}$ (e) $\frac{3}{10}$

Sections C and D (p 22)

1 (a) $\frac{2}{3}$ (b) $\frac{3}{7}$ (c) $\frac{5}{6}$

2 (a) $\frac{6}{5}$ (b) $\frac{11}{8}$ (c) $\frac{7}{3}$
   (d) $\frac{23}{8}$ (e) $\frac{13}{4}$

3 (a) $1\frac{1}{2}$ (b) $2\frac{2}{5}$ (c) $3\frac{3}{4}$
   (d) $2\frac{2}{3}$ (e) $3\frac{1}{3}$

4 $\frac{45}{20}$, $\frac{36}{15}$, $\frac{29}{12}$

Sections E and F (p 23)

1 (a) $\frac{9}{20}$ (b) $\frac{23}{24}$ (c) $\frac{17}{21}$

   (d) $\frac{31}{40}$ (e) $1\frac{1}{12}$ (f) $1\frac{1}{10}$

   (g) $1\frac{11}{24}$ (h) $1\frac{1}{12}$

2 (a) $\frac{7}{20}$ (b) $\frac{1}{56}$ (c) $\frac{1}{10}$

   (d) $\frac{13}{20}$ (e) $\frac{3}{14}$ (f) $\frac{13}{18}$

   (g) $\frac{5}{9}$ (h) $\frac{11}{14}$

3 (a) $3\frac{1}{2}$ (b) $3\frac{1}{3}$ (c) $7\frac{1}{2}$

4 $\frac{1}{10} + \frac{1}{6} = \frac{4}{15}$ $\frac{1}{10} + \frac{1}{4} = \frac{7}{20}$ $\frac{1}{6} + \frac{1}{4} = \frac{5}{12}$

   $\frac{1}{10} + \frac{1}{3} + \frac{13}{30}$ $\frac{1}{6} + \frac{1}{3} = \frac{1}{2}$ $\frac{1}{4} + \frac{1}{3} = \frac{7}{12}$

5 (a) $\frac{5}{2} + \frac{4}{3} = \frac{23}{6}(= 3\frac{5}{6})$

   (b) $\frac{2}{4} + \frac{3}{5} = \frac{11}{10}(= 1\frac{1}{10})$

6 (a) $\frac{8}{3} + \frac{6}{5} = \frac{58}{15}(= 3\frac{13}{15})$

   (b) $\frac{3}{6} + \frac{5}{8} = \frac{9}{8}(= 1\frac{1}{8})$

7 (a) $\frac{5}{2} - \frac{3}{4} = \frac{7}{4}(= 1\frac{3}{4})$

   (b) $\frac{3}{5} - \frac{2}{4} = \frac{1}{10}$

8 (a) $\frac{8}{3} - \frac{5}{6} = \frac{11}{6}(= 1\frac{5}{6})$

   (b) $\frac{5}{8} - \frac{3}{6} = \frac{1}{8}$

9 $4 \times \frac{5}{6}$ or $5 \times \frac{4}{6} = \frac{10}{3}(= 3\frac{1}{3})$

Review 1 (p 78)

1 (a) (b) (c)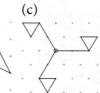

2 (a) 17 (b) 14

3 (a) 81 (b) B, C

4 (a) Subtract 7; ⁻4, ⁻11

  (b) Add 4; ⁻3, 5, 9, 17, 25

5 $a = 53°$, $b = 73°$, $c = 54°$, $d = 53°$, $e = 73°$, $f = 54°$

6 (a) $\frac{2}{5}, \frac{5}{12}, \frac{9}{20} \left(\frac{24}{60}, \frac{25}{60}, \frac{27}{60}\right)$

  (b) (i) $\frac{5}{12}$ (ii) $1\frac{19}{30}$ (iii) $\frac{23}{24}$

Mixed questions 1 (Practice booklet p 24)

1 (a) or

  (b) (c)

2 (a) 792 cans (b) 18 boxes

3 (a) 13, 34, 55, **76**, **97**, **118**

  (b) 16, **25**, 34, **43**, **52**, 61

  (c) **64**, 53, **42**, 31, **20**, 9

  (d) 18, **23**, **28**, 33, **38**, **43**

  (e) 17, **20**, **23**, **26**, 29, **32**

  (f) 40, **33**, **26**, 19, **12**, **5**

4 $a = 102°$ $b = 60°$ $c = 64°$
  $d = 72°$ $e = 27°$

5 The pupil's accurate drawing
  AC = 6.2 cm, BC = 6.0 cm

6 (a) $\frac{9}{40}$ (b) $2\frac{7}{12}$ (c) $1\frac{11}{30}$ (d) $\frac{41}{60}$

7 (a) (i) 12 (ii) 6 (iii) 3

  (b)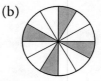

  (c) (i) 12 (ii) 6 (iii) 3

8 $\frac{13}{60}$

9 (a) 100

  (b) ▽ , because 400 is at the end of the 20th row and is in △ .

10 (a) $364 \div 14 = 26$

  (b) $23 \times 17 + 10 = 401$

Action and result puzzles (p 79)

In each puzzle, the action cards show operations to be performed on a starting number and the result cards show the results. Pupils match up the results with the actions.

The puzzles provide number practice and an opportunity to apply some logical thinking. They may reveal misconceptions about number.

Essential	Optional
Puzzles on sheets 108 and 109	Sheet 110 (blank cards)
Scissors	Puzzles on sheet 107 (easier)
	OHP transparencies of some sheets, cut into puzzle cards

'I put copies of the games into labelled boxes. I explained what each game was about and allowed them to choose their own games.'

A selection from the puzzles should be used (listed below roughly in order of difficulty).

Sheet 108   16.5 puzzle (+, −, × and ÷, decimals and negative numbers)

s puzzle (the number *s* has to be found)

12 puzzle (*a*, *b*, *c* and *d* are numbers to be found)

Sheet 109   8 puzzle (× and ÷, including fractions, simplifying fractions)

20 puzzle (*a*, *b*, *c* and *d* are numbers to be found)

Pupils who need a gentler start could do the following.

Sheet 107   36 puzzle (+, −, × and ÷, simple decimals and negative numbers)

q puzzle (+ and −, two-digit numbers, *q* has to be found)

h puzzle (+, −, × and ÷, two-digit numbers, *h* has to be found)

◊ This has worked well with pupils sitting in pairs at tables of four. When each pair had matched the cards, all four pupils discussed what they had done. An aim is to encourage mental number work. However, pupils may want to do some calculations and demonstrate things to their group using pencil and paper. A calculator should not be used.

'I copied the cards on to pieces of acetate which could be moved about on the OHP. Pupils went to the OHP to show how the cards matched up.'

◊ Puzzles that pupils find easy can be done without cutting out the cards: they simply key each action card to its result card by marking both with the same letter. However, something may be learnt from moving cards around to try ideas out before reaching a final pairing, and some puzzles are almost impossible unless they are done this way.

◊ Solutions can be recorded by
 • keying cards to one another with letters as described above
 • sticking pairs of cards on sheets or in exercise books
 • writing appropriate statements

◊ After pupils have solved some puzzles, they can make up some of their own (using the blank cards) to try on a partner. This may tell you something about the limits of the mathematics they feel confident with. Most should be able to make up puzzles of the *s* and 12 types.

◊ **Dividing by numbers between 0 and 1**

Some pupils were successful at reaching understanding by themselves and went on to explain division by fractions to other pupils. They thought about division as 'how many there are in' and reasoned, for example, from 'there are four 2s in 8' to 'there are sixteen halves in 8'.

Sheet 108

16.5 puzzle

Action	Result
+ 3.5	20
÷ 10	1.65
− 17	⁻0.5
× ⁻2	⁻33
÷ 0.5	33
+ ⁻3.5	13
− 6.05	10.45
× 0.1	1.65

s puzzle: $s = 4.5$

12 puzzle: $a = 0.5$, $b = 4$ and $c = 2$
(or $b = 2$ and $c = 4$), $d = 10$

Sheet 109

8 puzzle

Action	Result	Action	Result
÷ $\frac{1}{2}$	16	× 2	16
÷ 4	2	÷ 12	$\frac{2}{3}$
× $\frac{1}{8}$	1	× $\frac{1}{10}$	$\frac{4}{5}$
÷ $\frac{1}{4}$	32	÷ $\frac{1}{8}$	64
÷ 5	$1\frac{3}{5}$	× 4	32
× 10	80	÷ 8	1
÷ 2	4	÷ 10	$\frac{4}{5}$
× $\frac{1}{2}$	4	× $\frac{1}{4}$	2

20 puzzle: $a = \frac{1}{2}$, $b = ⁻20$, $c = 0.1$, $d = 40$

Sheet 107 (gentler start)

36 puzzle

Action	Result
÷ 9	4
− 40	⁻4
× 3	108
+ 27	63
÷ 8	4.5
× 1.5	54
+ ⁻50	⁻14
÷ 24	1.5

q puzzle: $q = 27$

h puzzle: $h = 9$

Chocolate (p 79)

This is a problem-solving activity. It gives pupils an opportunity to compare fractions, but is presented in such a way that they have to think out an approach for themselves.

Essential	**Optional**
Bars or blocks of something which can be divided up and shared out equally	Bars of chocolate (of a kind not already subdivided into portions)

◊ The problems are all variations of this basic idea:
  • A number of tables are set out, each with some chocolate bars.
  • A group of pupils are asked, one by one, to choose a table to sit at.
  • When everyone has sat at a table, the bars on each table are shared equally between those at that table.

The problem for the pupils is to decide which table to sit at in the hope of getting the most chocolate at the shareout.

Getting started

◊ It is best to start with a fairly simple situation, for example three tables with 1, 2 and 3 bars.

Choose a group of pupils to take part, say eight, and explain the problem. Ask them one by one to choose their table (they cannot change their minds later).

◊ As pupils choose where to sit, involve the whole class and ask questions such as:
  • Where would you sit? Why?
  • How much chocolate would each person get at this table if no one else sits here?
  • Is it best to be the first to choose, the last, or doesn't it matter?

◊ Once the last pupil has chosen, ask pupils to decide who gets the most chocolate and to justify their answer. Pupils could consider this in small groups and then give their explanations to the whole class.

Explanations could involve fractions, percentages, decimals, ratios, or they may be idiosyncratic (for example, giving each bar a particular weight and dividing).

Variations

◊ Obviously the number of tables and/or bars can be varied.

You could also tell pupils that you will decide beforehand how many pupils will sit down but they will not know this until you stop them and they then share out the chocolate.

Follow-up work

◊ Pupils could work in pairs or small groups on particular problems, such as

• Is it always the best strategy for the first person to sit at the table with most chocolate?

For example, suppose there are three pupils and three tables with 1, 2 and 3 bars.

Suppose the first pupil goes to the table with 3 bars.
The second pupil's best choice appears to be the table with 2 bars.
The third pupil's best choice is the table with 3 bars.

The second pupil has done best. So the first pupil would have done better to have gone to this table.

11 Health club

This practical introduction to some of the ideas of data handling is based on a set of data cards. The cards can be sorted, ordered, arranged to make bar charts, and so on.

A lot of ideas are introduced in this unit. They are all taken up again in more detail later in the course.

p 80	**A** On record	Using information on data cards
p 80	**B** On display	Median (odd number of values) Dot plots, grouped bar charts
p 82	**C** Males and females	Median (even number of values) Comparing two sets of data
p 82	**D** Two-way tables	
p 82	**E** Healthy or not?	Using a formula for body mass index

---

Essential

Sets of cards (one per pair of pupils) from sheets 80 and 81
Sheets 82, 83, 86, 87 (unless pupils draw their own axes for graphs)

Practice booklet pages 26 and 27

A On record (p 80)

This is to familiarise pupils with the data cards.

> A set of cards per pair of pupils (from sheets 80 and 81)

◊ You could start by asking pupils why a health club would keep the kind of information on the cards (which would of course have to be updated periodically). Explain that rest pulse (in beats per minute) is measured when the person is sitting still. It is a rough measure of fitness (the lower the better).

◊ Ask some questions to help familiarise pupils with the information on the cards, such as 'Who weighs 60 kg?' or 'Who is the tallest female?'

B On display (p 80)

This introduces grouped frequency bar charts, the median and dot plots.

> The cards, sheets 82, 83 (unless pupils draw their own axes for graphs)

◊ Ask pupils, in pairs, to sort the cards by age group. The cards can be placed to form a 'bar chart'. (You can do this on the board with Blu-tack.)

'*I used a demo set of plastic coated cards. Pupils enjoyed putting them into grouped bar charts or on dot plots. Each child came and put a card on the board.*'

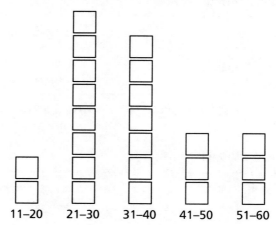

11–20   21–30   31–40   41–50   51–60

Pupils then draw the bar chart (on sheet 82, if used).

Discuss the distribution of age groups and the possible reasons for this.

Median

◊ Ask pupils to look at the heights on the cards, to put the cards in order of height and to find the middle height. (Some pupils may think of this as halfway between the heights of the shortest and tallest people. You can say that this is one interpretation of 'middle height' but not what you meant.)

Emphasise that the median is a height (168 cm), not a person. (It helps to use 'median' always as an adjective before 'height', 'weight', etc.) Emphasise also that there are equal numbers of people shorter than and taller than the median height.

Dot plot

◊ The cards themselves can be used to illustrate a dot plot of the heights. Place them on a large number line using Blu-tack:

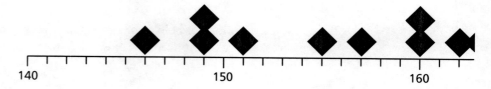

140   150   160

Pupils draw the dot plot on sheet 82.

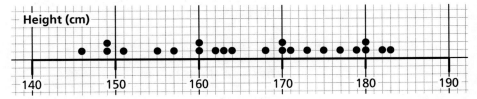

Height (cm)

They could now discuss what the dot plot shows about the heights: they are fairly evenly spread but with more at the taller end.

Ask what sort of information can quickly be found from looking at the dot plot rather than the cards (for example, that the tallest person is 183 cm).

Grouped bar chart

◊ Compare the dot plot for heights with the bar chart for ages. The dot plot gives more detail but the bar chart shows the overall shape of the distribution better.

Discuss how to draw a bar chart for heights by grouping them.

Pupils draw a grouped bar chart on sheet 82. They can work directly from the cards or use the dot plot.

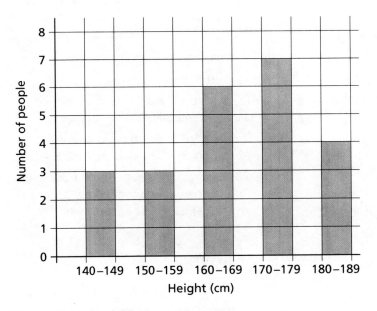

The bar chart shows clearly that most people at the club are between 160 and 179 cm tall.

Some pupils could look at the effect of choosing different class intervals.

B5   This raises the issue of the median of an even number of values.

C Males and females (p 82)

This is about comparing two sets of data.

The cards, sheets 86 and 87

Further work

The cards can also be used to show a scatter diagram. For example, you could draw axes on the board for height and weight and Blu-tack each card at its position in the scatter diagram.

D Two-way tables (p 82)

The cards

◊ The cards can be sorted into a large two-way diagram for example by age and gender:

	Age	
	11–40	**41–70**
Male	■ …	■ ■ …
Female	■ ■ ■ …	■ …

From this a two-way table of numbers can be made.

In one school, after working through the unit, pupils had a further lesson using a spreadsheet. The teacher had entered the data from the cards.

Pupils were shown how to sort the data, for example in ascending order of rest pulse. They also used the median command.

They were shown how to filter the data to select out all the females and find median values for this subgroup. They then tackled questions C1(a) and C2(a) using the spreadsheet.

E Healthy or not? (p 82)

This uses the formula for 'body mass index'.

The cards

◊ Emphasise that the formula is valid only for adults. You may need to be sensitive to pupils' concerns about weight.

Body mass index is a crude measure and has been criticised by some medical authorities as misleading since it takes no account of bone, muscle, etc. as a proportion of mass.

Ⓐ On record (p 80)

A1 (a) 39 kg

(b) J Abram

(c) 8

(d) A Saunders, W Stanwell

(e) 13

(f) H Tear

(g) 5

(h) 2

(i) M Lakhani, G Peters

Ⓑ On display (p 80)

B1 (a) 64 kg

(b)

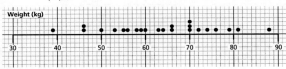

(c)

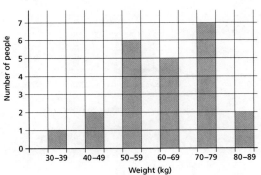

(d) The pupil's observations: for example, most people at the club weigh between 50 kg and 80 kg and not many people weigh over 80 kg.

B2 (a) 69 bpm

(b)

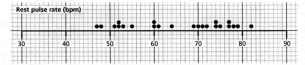

(c)

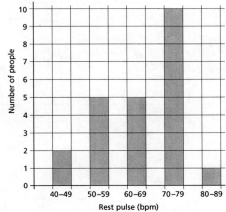

(d) The pupil's observations: for example, almost half of the people at the club have a rest pulse rate between 70 and 79 bpm.

B3 (a) The actual ages are not recorded.

(b) 32 years and 39 years are possible median ages.

B4 22 years and 25 years are possible median ages.

B5 (a) 167 cm

(b) 62.5 kg or $62\frac{1}{2}$ kg

ℂ **Males and females** (p 82)

C1 (a) Males 70 kg, Females 57 kg

(b) The median weight for the males is 13 kg higher than the median weight for the females suggesting that the males at the health club tend to be heavier than the females.

(c)

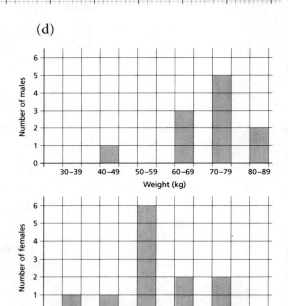

(d)

(e) The dot plots and bar charts support the claim that the men at the club tend to be heavier than the women. The dots for the males are further over to the right. On the bar chart for the males, the bars tend to be higher over to the right. The charts show that all the males except one weigh 60 kg or over but 8 out of the 12 females weigh below 60 kg.

C2 (a) The median rest pulse rate for the males is 69 bpm.
The median rest pulse rate for the females is 65 bpm.

(b) The median rest pulse rate for the males is only 4 bpm higher than the median rest pulse rate for the females suggesting that there is little difference between the rest pulse rates of the males and females at the health club.

(c)

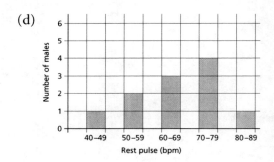

(d)

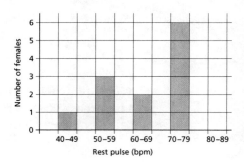

(e) The medians do not show a marked difference between the rest pulses for the males and females at the club. The dot plots show a fairly even spread for the males but a less even spread for the females where the dots tend to be clustered round two points. The bar chart for the females shows a 'dip' in the middle not shared by the graph for the males.

D Two-way tables (p 82)

D1

	Less than 150 cm	150 cm or more
Less than 50 kg	1	2
50 kg or more	2	18

D2

	Less than 60 bpm	60 bpm or more
11–40	3	13
41–70	4	3

E Healthy or not? (p 82)

E1 Members of the club in the 11–20 age group are excluded as they may not be adults.

	Over a healthy weight	Under a healthy weight
Males	J Abram, S Anderson, L Unwin	
Females	W Evans, E Pransch, K Quaraishi	J Dellano, W Stanwell

Practice booklet

Sections B and C (p 27)

1 (a)

(b) 1.80 m

2 (a)

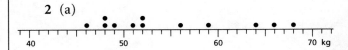

(b) 52 kg

3 There are 25 members altogether. The median age is that of the 13th member in order.
There are 13 members altogether in the 21–30 and 31–40 groups so the 13th is in the **31–40** group.

Section D (p 27)

1 (a)

	Swimmers	Non-swimmers
Men	**10**	**3**
Women	**6**	**6**

(b)

Age	Swimmers	Non-swimmers
21–40	9	4
41–60	7	5

(c)

Age	Men	Women
21–40	6	7
41–60	7	5

⑫ Balancing

This work introduces, in an informal way, the idea of doing the same thing to both sides of an equation.

T	p 83 **A** Scales	Introduction to the idea of 'doing the same thing' to both sides of a balance
T	p 84 **B** Writing	Writing and solving a balance puzzle using a letter, for example $n + n + n = n + 10$
	p 86 **C** Using shorthand	Using letters to help solve puzzles Using $3n$ notation instead of $n + n + n$

Practice booklet pages 28 and 29

Ⓐ **Scales** (p 83)

This section introduces the idea of scales balancing. The emphasis is on what can be done to both sides and still leave the scales in balance.

Use the pictures to discuss how to find out what one of each object weighs, by taking things from both sides of the scales. Point out that all the weighs are (notionally) 1 kilogram each.

You may wish to do the questions orally. Pupils are not expected to show any working when solving these puzzles. Of course, if they wish to write down any intermediate steps they should not be discouraged from doing so.

Ⓑ **Writing** (p 84)

◊ When introducing this section to pupils, point out that we are simply recording what is done to the balance in the most straightforward way possible.

Emphasise that the unknown here (n in the example) stands for the *number of kilograms* each animal weighs. Avoid using letters like h (which pupils may think stands for the word 'hedgehog') or w (which pupils may think of as standing for 'a weight').

B3 Pupils choose their own letters to stand for the weight of each animal. Emphasise that the letter stands for the weight of the animal, not the name of the animal or the animal itself.

C Using shorthand (p 86)

This section encourages higher attainers to use the notation $3n$ as a shorthand for $n + n + n$ when solving problems.

This is the first time that pupils have to take both weights and unknown objects from each side. The emphasis is on writing out the solution to the puzzle step by step.

A Scales (p 83)

A1 (a) 2   (b) 3   (c) $2\frac{1}{2}$ (d) 5
     (e) 3   (f) 12   (g) $1\frac{1}{2}$ (h) $12\frac{1}{2}$
     (i) 21

B Writing (p 84)

In this and the following sections, pupils should write down their working and check.

B1 $n = 4$

B2 $x = 3$

B3 (a) 5   (b) 4   (c) 11   (d) 20

C Using shorthand (p 86)

C1 The pupil's check that $n = 2$ fits the puzzle

C2 (a) 4   (b) 3   (c) 2

C3 (a) 8   (b) 79

C4 The pupil's picture for $3h + 15 = h + 37$, solution $h = 11$

C5 (a) $p = 7$ (b) $d = 6$ (c) $s = 8$
     (d) $q = 10$ (e) $n = 4$ (f) $u = 53$
     (g) $k = 2\frac{1}{2}$ (h) $y = \frac{1}{3}$ (i) $a = 5$
     (j) $w = \frac{1}{2}$ (k) $b = 20$ (l) $f = 3.2$
     (m) $j = 17$ (n) $s = 2$

C6 (a) $d = 10$ (b) $e = 6$ (c) $k = 0.9$
     (d) $y = 3.5$

What progress have you made? (p 87)

1 $2n + 11\frac{1}{2} = 4n + 7\frac{1}{2}$
     $n = 2$

2 (a) $y = 4$ (b) $h = 13$ (c) $f = 5$
     (d) $w = 12$

Practice booklet

Section B (p 28)

  1 3

  2 0.5

  3 4

  4 50

  5 12

  6 15

  7 37

Section C (p 29)

  1 (a) $3w + 9 = 16 + w$ (b) $w = 3.5$

  2 (a) $t = 6.5$ (b) $w = 6$ (c) $m = 12$
      (d) $t = 16$ (e) $p = 11$ (f) $x = 10$

  3 (a) $t = 100$ (b) $w = 0.6$ (c) $m = 0.12$
      (d) $t = 160$ (e) $p = 110$ (f) $x = 1000$

  4 $7c + 3.9 = 23.4 + c$
     $c = 3.25$

⑬ Multiples and factors

Essential

Sheets 134 to 138
Scissors

Practice booklet pages 30 and 31

A The sieve of Eratosthenes (p 88)

> Sheet 134

◊ Eratosthenes of Cyrene (276–194 BCE) was a Greek who is best known for measuring the Earth's circumference by observing the direction of the Sun at two places a great distance apart.

The 'sieve' is his best-known contribution to mathematics. The ringed numbers are, of course, the prime numbers.

B Factor pairs (p 88)

C Factor trees (p 89)

◊ If we allow, 2 to lead to the pair '2, 1' then the tree will go on for ever.

◊ It is common for pupils to forget about multiplication and write down pairs which add together to make the number.

D Prime factorisation (p 90)

E Lowest common multiple (p 91)

F Highest common factor (p 92)

G Testing for prime numbers (p 93)

Pupils should have all the prime numbers less than 100 from the sieve of Eratosthenes.

H Clue-sharing (p 94)

> Sheets 135 to 138, scissors

◊ These puzzles are designed to encourage pupils to work collaboratively in pairs or small groups. The digits are to be cut out and put in the squares according to the clues. The clue cards are dealt out to the group members. Only the person who gets a card is allowed to see it, so they have to tell the others the clues on their cards. This makes it impossible for anyone to 'opt out', as all the clues are needed to solve the puzzle.

◊ Challenge pupils to find which puzzle has redundant clues ('Star'). Which clues are redundant? Why?

◊ After solving some of the puzzles, pupils could have a go at making up their own puzzles with clue cards.

◊ The solutions to the four puzzles are as follows.

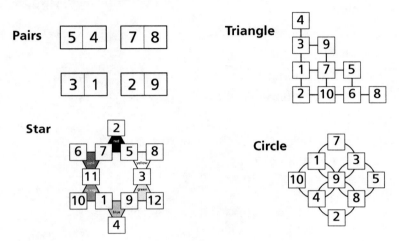

Problems (p 94)

Factor pairs (p 88)

B1 (a) 1 2 3 4 6 12

(b) 1 2 3 6 9 18

(c) 1 2 3 4 6 8 12 24

(d) 1 2 3 5 6 10 15 30

B2 (a) 1 2 3 4 6 9 12 18 36

There is an odd number of factors. One factor pairs off with itself.

(b) Square numbers (e.g. 4, 9, 16, ...)

B3 (a) 1 ⌣ 17

There are only two factors, 1 and the number itself, because 17 is prime.

B4 When 10 is a factor: number ends in 0.

When 5 is a factor: number ends in 5 or 0.

When 4 is a factor: last two digits make a number which is a multiple of 4.

When 6 is a factor: number is even and 3 is a factor (using the test for 3).

When 9 is a factor, the sum of the digits is a multiple of 9.

C Factor trees (p 89)

C1 (a), (b)

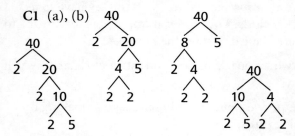

The trees all end with the same set of numbers, 2, 2, 2, 5.

C2 (a) Tree ending 2, 2, 7
   (b) Tree ending 2, 3, 5
   (c) Tree ending 2, 2, 2, 3, 3
   (d) Tree ending 2, 2, 5, 5

C3 (a) A = 90, B = 6, C = 15
   (b) P = 1620, Q = 54, R = 30, S = 6, T = 9, U = 15
   (c) V = 18, W = 10, X = 9, Y = 2, Z = 3
       or
       V = 12, W = 15, X = 4, Y = 3, Z = 2
   (d) J = 36, K = 3, L = 6, M = 2

D Prime factorisation (p 90)

D1 (a) $2 \times 2 \times 2 \times 5$ (b) $2 \times 2 \times 7$
   (c) $2 \times 3 \times 5$ (d) $2 \times 2 \times 2 \times 3 \times 3$
   (e) $2 \times 2 \times 5 \times 5$

D2 (a) 5^3 (b) 3^5
   (c) 7^3 (d) $2^3 \times 3^2$
   (e) $3^2 \times 5^3 \times 7^2$

D3 (a) 2^4 (b) 2×5^2
   (c) $2^4 \times 3$ (d) $2^4 \times 5$
   (e) $2^2 \times 3 \times 7$ (f) 2×3^3

D4 (a) $2^6 \times 3$ (b) 5×11^2
   (c) $3^3 \times 5^2$ (d) $2^3 \times 3^2 \times 13$

E Lowest common multiple (p 91)

E1 (a) 4, 8, 12, 16, 20, 24, ...;
       6, 12, 18, 24, 30, 36, ...
   (b) 12

E2 (a) 40   (b) 60   (c) 60   (d) 60

E3 36

E4 $2 \times 2 \times 3 \times 7 = 84$

E5 $18 = 2 \times 3 \times 3$; $45 = 3 \times 3 \times 5$
   LCM $= 2 \times 3 \times 3 \times 5 = 90$

E6 (a) 150   (b) 240   (c) 48
   (d) 126   (e) 168

F Highest common factor (p 92)

F1 (a) 1, 2, 4, 8, 16; 1, 2, 4, 5, 10, 20
   (b) 4

F2 (a) 4   (b) 10   (c) 15   (d) 2

F3 4

F4 $2 \times 3 \times 3 = 18$

F5 $32 = 2 \times 2 \times 2 \times 2 \times 2$
   $80 = 2 \times 2 \times 2 \times 2 \times 5$
   HCF $= 2 \times 2 \times 2 \times 2 = 16$

F6 (a) 15   (b) 16   (c) 8
   (d) 6   (e) 1

F7 (a) 240   (b) 15   (c) 6   (d) 144

F8 (a) 180   (b) 12   (c) 14   (d) 30

F9 (a) 240   (b) 12

G Testing for prime numbers (p 93)

G1 1927 is not prime; it is divisible by 41.

G2 1931 is prime.
   As you try each prime number, the other 'factor' gets smaller. When the two are as close as possible then there is no need to try any further.

G3 The statement is true for $n =$ 1, 2, 3, 4, 5, 6, ... up to 15.
   But it is false for $n = 16$
   ($16^2 + 16 + 17 = 17^2$)
   and obviously false for $n = 17$.

G4 $(2 \times 3 \times 5 \times 7) + 1 = 211$, which is prime.
   $(2 \times 3 \times 5 \times 7 \times 11) + 1 = 2311$, which is prime.

$(2 \times 3 \times 5 \times 7 \times 11 \times 13) + 1 =$
30 031, which is divisible by 59.

$(2 \times 3 \times 5 \times 7 \times 11 \times 13 \times 17) + 1 =$
510 511, which is divisible by 19.

▯ Problems (p 94)

I1 360 seconds

I2 $240 = 2 \times 2 \times 2 \times 2 \times 3 \times 5$

Any combination picked from this
collection will give a factor of 240,
e.g. $2 \times 2 \times 5$.

The complete set (apart from 1) is
2, 2×2, $2 \times 2 \times 2$, $2 \times 2 \times 2 \times 2$,

then each of these with $\times 3$, with $\times 5$,
and with $\times 3 \times 5$, and 3, 5 and 3×5.

There are 20 factors altogether.

*I3 (a) 61 (add 1 to the LCM of 2, 3, 4, 5,
and 6)

(b) 59 (The number which is 1 less than
the LCM of 2, 3, 4, 5 and 6 will leave
a remainder of 1 when divided by 2,
of 2 when divided by 3, etc.)

The numbers are spaced at intervals
of 60. There are 16 of them below
1000 (59, 119, 179, … , 959).

The problems can also be approached as
follows

Numbers leaving remainder 1 when
divided by 2: 1, 3, 5, 7, 9, … and so on

What progress have you made? (p 94)

1 (a) Tree ending 3, 3, 5

(b) Tree ending 2, 2, 3, 5

2 (a) $2 \times 2 \times 2 \times 2 \times 2 \times 3$

(b) $2 \times 2 \times 3 \times 5 \times 11 \times 13$

3 $2^5 \times 3$

4 56

5 15

Practice booklet

Sections B, C and D (p 30)

1 (a) 7 + 23, 11 + 19, 13 + 17

(b) 3 + 97, 11 + 89, 17 + 83, 29 + 71,
41 + 59, 47 + 53

2 (a) Tree ending 2, 2, 3, 3

(b) Tree ending 2, 5, 7

3 (a) A is 15, B is 30, C is 6 and D is 180.

(b) A is 2, B is 60, C is 30, D is 15,
E and F are 3 and 5.

4 (a) $2^2 \times 3 \times 5$

(b) $600 = 60 \times 2 \times 5 = 2^3 \times 3 \times 5^2$

5 $2^6 \times 3^2 \times 5^2$

6 (a) $a = 3$, $b = 2$ (b) $a = 2$, $b = 3$, $c = 7$

7 (a) 2^4 (b) $2^2 \times 3^2$

(c) $2^2 \times 5^2$ (d) $2^2 \times 3^2 \times 5^2$

(e) In the prime factorisation of a square
number every prime factor is raised
to an even power.

8 (a) $2^2 \times 3^2 \times 7 \times 23$ (b) $2 \times 3^5 \times 11$

(c) $2^4 \times 3^2 \times 53$

Sections E, F, G and I (p 31)

1 (a) 60   (b) 84

2 (a) 108   (b) 360   (c) 22 050   (d) 8100

3 24

4 (a) 6   (b) 10

(c) $3^2 \times 7 = 63$ (d) $2^2 \times 5^2 = 100$

5 23

6 They will ring again when the time passed
in minutes is equal to the LCM, that is 288
minutes later, or 4:48 p.m.

7 The jug size will be the HCF of 72, 24, 56
and 120.
This is 8 litres.

8 101, 131 and 151

Work to rule

The emphasis in this unit is on finding rules by analysing tile designs.

p 95	**A** Tile designs	Finding and explaining a rule to calculate the number of white tiles given the number of red tiles
p 98	**B** Shorthand	Understanding algebraic shorthand: $w = 2r + 1$
p 99	**C** Using shorthand	Finding and using rules such as $w = 4r + 3$
p 101	**D** T-shapes and more	Finding and using rules such as $w = \dfrac{r}{2} + 3$
p 103	**E** Snails and hats	Considering designs that give quadratic rules

---

Optional
Sheet 89

Practice booklet pages 32 and 33

𝔸 **Tile designs** (p 95)

In this section pupils should become aware that there are two types of rule for these patterns:

- by tabulating results in order we can see how the sequence of white tiles continues – it goes up in 2s
- by looking at the structure of the pieces we can see that the number of whites is equal to double the number of reds plus 3

Pupils should begin to appreciate the advantages of the latter rule.

◊ You can start by discussing the design of the pieces in the mobile on page 97, perhaps using tiles or multilink.

Tabulate results in order up to, say, 8 red tiles, and look at the pattern in the table. Many pupils will spot that the number of white tiles goes up in 2s, but they should think about why it will continue in 2s.

Ask pupils to imagine the piece with, say, 100 red tiles and how we could calculate the number of white tiles. Pupils should appreciate it would take a long time to continue to add on 2s.

A discussion of the structure of the designs should lead to
- the piece with 100 red tiles has (100 × 2) + 3 whites

and to the general rule

number of whites = (number of reds × 2) + 3

Use the discussion to bring out how diagrams can be useful in making the structure of the designs clear, for example:

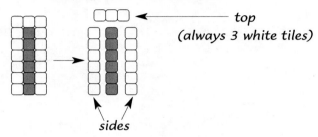

B Shorthand (p 98)

◊ The convention that multiplication has priority over addition is not used here. Brackets are used throughout.

◊ You may need to emphasise that *2r* means '2 × the number of red tiles', not '2 red tiles'.

C Using shorthand (p 99)

C7 This is intended to provide pupils with an opportunity to structure their own work to find rules. Encourage them to use strategies such as counting tiles in the examples given, drawing more diagrams and tabulating results to help them analyse the diagrams to find a rule.

Pupils could work in groups and present their ideas on one or more of these designs to the whole class.

D T-shapes and more (p 101)

D1 Pupils can select one or more of these designs to investigate. They could work in groups and present their ideas to the whole class.

E Snails and hats (p 103)

Optional: Sheet 89

*J18 This question is on sheet 89 and will extend even the most able.

A Tile designs (p 95)

A1 (a)

Number of red tiles	1	2	3	4	5	6
Number of white tiles	3	5	7	9	11	13

(b) (i) 41 white tiles

(ii) 201 white tiles

(c) number of white tiles =
(number of red tiles) × 2 + 1

(d) 145 white tiles

A2 (a) The pupil's response

(b) The pupil's response

A3 (a) The pupil's response

(b) The pupil's response and explanation

A4 number of white tiles =
(number of red tiles) × 2 + 6

B Shorthand (p 98)

B1 A and F, B and H, C and E, D and G

B2 Any letters can be used to stand for the
number of white tiles and the *number of
red tiles*. We have used *w* for the *number
of white tiles* and *r* for the *number of red
tiles*.

(a) $w = 4r + 3$ or equivalent

(b) $w = 5r$ or equivalent

(c) $w = 2r + 7$ or equivalent

(d) $w = r + 6$

(e) $w = 3r - 1$ or equivalent

C Using shorthand (p 99)

C1 (a) 23 white tiles

(b) The pupil's description of a check,
possibly involving drawing the piece
with 5 tiles

C2 (a) 51 white tiles

(b) 243 white tiles

C3 (a) Any rule equivalent to
number of white tiles =
(number of red tiles × 4) + 3

(b) $w = 4r + 3$ or equivalent

(c) The pupil's explanation of why the
rule works

C4 (a) 103 white tiles

(b) 163 white tiles

C5 (a) The pupil's design that fits
$w = 2r + 2$

(b) The pupil's explanation (with
diagrams)

C6 (a) $w = r + 1$

(b) The pupil's explanation of why the
rule works

C7 For each rule (or its equivalent) the pupil
should give an explanation of why it
works.
Set A: $w = r + 4$
Set B: $w = 4r + 10$
Set C: $w = 3r + 1$
Set D: $w = 5r + 3$

D T-shapes and more (p 101)

D1 For each rule (or its equivalent) the pupil
should give an explanation of why it
works.

Set A: (a) $w = \frac{r}{2} + 6$ (c) 56 white tiles

Set B: (a) $w = r - 2$ (c) 98 white tiles

Set C: (a) $w = \frac{r}{2} + 3$ (c) 53 white tiles

Set D: (a) $w = r + 4$ (c) 104 white tiles

Set E: (a) $w = 2r - 4$ (c) 196 white tiles

Set F: (a) $w = \frac{3r}{2} + 5$ (c) 155 white tiles

E Snails and hats (p 103)

E1 37 white tiles

E2 The pupil's explanation of the rule
$w = (r \times r) + 1$

E3 12 red tiles

E4 100 white tiles

E5 The pupil's explanation of the rule
$w = (r - 2) \times (r - 2)$

E6 529 white tiles

E7 10 red tiles

E8 200 is not a square number

*E9 (Sheet 89)

A (a) 102 white tiles

(b) 10 002 white tiles

(c) $w = g^2 + 2$

(d) 2502 white tiles

B Pupils may find equivalents to these answers.

Set A: $w = \left(\frac{g}{4}\right)^2 + 4$

Set B: $w = g^2 + 2g + 4$

Set C: $w = \left(\frac{g-1}{2}\right)^2$

(g has to be an odd number.)

Set D: $w = g(g-1)$ or $w = g^2 - g$

Set E: $w = 3g + 4$

What progress have you made? (p 104)

1 (a) $w = r + 8$

(b) $w = 4r + 1$ or $w = (r \times 4) + 1$

2 (a) $w = 2r + 4$ or equivalent

(b) The pupil's explanation

(c) 104 white tiles

3 Set X $3r \div 2$

(a) $w = r - 1$ or equivalent

(b) The pupil's explanation

(c) 99 white tiles

Set Y

(a) $w = \frac{r}{2} - 2$ or equivalent

(b) The pupil's explanation

(c) 48 white tiles

Practice booklet

Section A (p 32)

1 The numbers of circles in the table are
1 3 5 7 9 11

2 $c = 2t - 1$, where c is the number of circles and t is the number of tiles.

3 119 circles

4 43 tiles

Sections B, C and D (p 32)

1 $c = 3t - 2$

2 $c = 1\frac{1}{2}t - 1$
There is one circle in the centre of each tile, plus a middle row of circles containing one less circle than the upper and lower rows.
So $c = t + \frac{1}{2}t - 1 = 1\frac{1}{2}t - 1$

3 (a) $c = 3t - 3$
There are 3 rows of circles; the number of circles in each row is one less than the number of tiles.

(b) $c = \frac{5t}{2} - 2$ or $c = 2\frac{1}{2}t - 2$
There are 2 rows of circles in which the number of circles is one less than the number of tiles; there is also a middle row on the joins in which the number of circles is half the number of tiles.

Section E (p 33)

1 $c = (t - 1)^2 + t^2$

Decimals 1

p 105 **A** Multiplying and dividing 10, 100, …

p 106 **B** Multiplying by 3, 0.3, 0.03, …

p 107 **C** Rounding decimals

p 107 **D** Dividing a decimal by a whole number

T p 108 **E** Rough estimates

p 109 **F** Problems

p 110 **G** Division: unit costs

p 112 **H** More problems

T p 113 **I** Dividing by numbers between 0 and 1

p 116 **J** Mixed questions

Practice booklet pages 34 to 39

Ⓐ **Multiplying and dividing 10, 100, …** (p 105)

◊ Emphasise place value: the figures move to the next higher place value when a number is multiplied by 10.

Ⓑ **Multiplying by 3, 0.3, 0.03, …** (p 106)

Ⓒ **Rounding decimals** (p 107)

◊ Pupils should gradually become familiar with different ways of describing the same degree of rounding: 'to one decimal place', 'to the nearest tenth', 'to the nearest 0.1'.

Ⓔ **Rough estimates** (p 108)

◊ A calculator is expected to be used from this section onwards, unless the question says otherwise.

◊ You can use calculations A–I for teacher-led discussion of methods of estimation. At this stage it is not expected that these will be formalised (for example, by expressing numbers to one significant figure). Any rough approximation will do if it will detect answers that cannot be right.

F Problems (p 109)

Target 100

This game offers a good reminder of the effect of multiplying by a number between 0 and 1.

G Division: unit costs (p 110)

◊ Allowing pupils to work in small groups with the potato problem gives them the chance to sort out the general method required as well as to discuss the need sometimes to round to the nearest penny.

Shop C has the cheapest potatoes (39p, to the nearest penny, per kg), then A (40p per kg), then B (45p per kg).

H More problems (p 112)

I Dividing by numbers between 0 and 1 (p 113)

From their experience with whole numbers, pupils may assume that dividing makes something smaller. They will find this is no longer true with numbers between 0 and 1.

The approach in this section is based on thinking about division as 'how many ... are there in ...?'

◊ You can help to show the meaning of each answer by rephrasing it as a multiplication. For example, $12 \div 0.5 = 24$ is equivalent to $24 \times 0.5 = 12$.

◊ Questions I1 to I3 can be done with the whole group working together.

◊ For questions I8 to I16 a fair amount of trial and error is expected. Afterwards you may want to bring out some general points (for example, the largest result comes from dividing the largest number by the smallest).

Target 20 (p 115)

This game focuses on the effect of dividing by a number between 0 and 1.

J Mixed questions (p 116)

Ⓐ Multiplying by 10, 100, ... (p 105)

A1 (a) 34　(b) 70　(c) 24.3
(d) 0.165　(e) 0.082　(f) 2.8
(g) 0.009　(h) 0.05

A2 (a) 10　(b) 100　(c) 100
(d) 100　(e) 100　(f) 10

A3 (a) 7400　(b) 520　(c) 0.0843
(d) 0.0034

A4 (a) 0.45　(b) 0.00042　(c) 0.0006
(d) 10 400

A5　$0.17 \times 1000 = 170$
$0.045 \times 1000 = 45$
$0.09 \times 1000 = 90$

A6　$670 \div 1000 = 0.67$
$6700 \div 1000 = 6.7$
$30 \div 1000 = 0.03$
$300 \div 1000 = 0.3$
$67 \div 1000 = 0.067$

Ⓑ Multiplying by 3, 0.3, 0.03, ... (p 106)

B1 £1.62

B2 £1.08

B3 £2.10

B4 (a) 29.4　(b) 13　(c) 12.48
(d) 138　(e) 7.76

B5 (a) 0.7　(b) 13.5　(c) 0.14
(d) 32　(e) 0.9

Ⓒ Rounding decimals (p 107)

C1 (a) 4.7　(b) 13.1　(c) 29.5
(d) 3.0　(e) 9.0

C2 (a) 2.88　(b) 8.98　(c) 1.40
(d) 14.70　(e) 0.10

C3 (a) 3.0　(b) 0.280　(c) 18.10
(d) 3.300

Ⓓ Dividing a decimal by a whole number (p 107)

D1 (a) 2.9　(b) 1.825　(c) 3.87
(d) 8.03　(e) 0.51

D2 (a) 0.134　(b) 0.68　(c) 0.1675
(d) 0.117　(e) 4.67

Ⓔ Rough estimates (p 108)

E1 Because it's about $3 \times 6 = 18$.
(He probably put a decimal point in the wrong place.)

E2 The answer is about $4 \times 0.8 = 3.2$, so 3 is a reasonable choice.

E3 (a) 20, 20.01　(b) 7 or 8, 7.38
(c) 10, 10.1　(d) 30, 28.3143

E4 (a) 14, 14.4　(b) 20, 19.74
(c) 5, 5.4027　(d) 16, 15.99
(e) 80, 81.9　(f) 90, 87.591

E5 (a) 21, 21.12　(b) 600, 696.6
(c) 24, 22.23　(d) 2.5, 2.62
(e) 0.4, 0.43　(f) 40, 39.29
(g) 0.2, 0.20　(h) 6.3, 6.26

Ⓕ Problems (p 109)

F1 Estimate $14 - 10 = 4$,
$14.3 \,\text{m} - 9.65 \,\text{m} = 4.65 \,\text{m}$

F2 Estimate $6 \times 4 = 24$,
5.5 square metres $\times 4 = 22$ square metres

F3 Estimate $1.2 - 1 = 0.2$,
$1.21 \,\text{m} - 0.95 \,\text{m} = 0.26 \,\text{m}$

F4 $£2.38 \times 4.5 = £10.71$
She needs 71p more.
(Estimating does not help here.)

F5 Estimate $0.6 \times 4 = 2.4$,
$0.63 \times £3.85 = £2.43$ to the nearest p.

F6 $8.45 \,\text{m} + 7.9 \,\text{m} = 16.35 \,\text{m}$, so it's not long enough. (Estimating by rounding to the nearest whole metre does not help.)

F7 Estimate 53 − (24 + 17) = 12,
52.5 kg − (23.66 kg + 17.48 kg) = 11.36 kg

F8 Estimate (0.8 × 7) + 1.2 = 6.8,
(0.78 kg × 7.25) + 1.22 kg = 6.875 kg

Ⓖ Division: unit costs (p 110)

G1 (a) 3 kg bag £1.42, 2 kg bag £1.45
(b) The 3 kg bag because 1 kg of onions in it costs less than in the 2 kg bag.

G2 (a) 4 kg bag £2.99, 2.5 kg bag £3.00
(b) The 4 kg bag, but only just because 1 kg of fuel in the 4 kg bag costs slightly less than in the 2.5 kg bag.

G3 (a) 40 litre bag 14p, 15 litre bag 15p
(b) The 40 litre bag because the cost per litre is lower.

G4 (a) 3.5 kg bag £2.84, 5 kg bag £2.75
(b) The 5 kg bag because the cost per kilogram is lower.

G5 (a) £2.51 per kilogram
(b) £1.52 per metre
(c) 87p per kilogram
(d) 44p per litre

G6 The 1.5 litre carton because it costs 73p per litre compared with 77p.

G7 The 1.8 kg packet because it costs 94p per kg compared with £1.15.

Ⓗ More problems (p 112)

H1 £1.80 ÷ 2.5 = 72p per litre

H2 £1.25 × 6.6 = £8.25

H3 (a) The 20 kg bag costs 26p per kg. The 15 kg bag costs 25p per kg. You get more for your money with the 15 kg bag.
(b) (2 × £3.75) + £5.20 = £12.70

H4 (a) 35.50 ÷ 6.99 = 5.07…
She can buy 5 CDs.
(b) £35.50 − (5 × £6.99) = £0.55 left over

H5 With the 400 g pot you pay 15p for 100 g.
With the 500 g pot you pay 16p for 100 g.
So the 400 g pot is better value.

*H6 (a) 9.4 + 7.7 = 17.1
(b) 31.65 + 59.66 = 91.31
(c) 8.4 − 3.9 = 4.5
(d) 4.5 × 8.7 = 39.15
(e) 70.03 − 49.4 = 20.63
(f) 2.3 × 5.6 = 12.88
(g) 6 × (4.8 + 5.5) = 61.8
or 6 × (4.8 + 6.5) = 67.8
(h) 3.5 × (2.3 + 1.1) = 11.9
or 3.5 × (1.3 + 1.9) = 11.2
(i) 9 × (8.2 − 4.7) = 31.5

Ⅰ Dividing by a number between 0 and 1 (p 113)

I1 Pupils are not expected to find the exact boundary beyond which the result will be over 200, and so on. However, to help check their results the boundaries are given here.

Over 200:
divide by numbers less than 0.06
Over 500:
divide by numbers less than 0.024
Over 1000:
divide by numbers less than 0.012

I2 To get a result between 2 and 3 you have to divide 20 by any number between 6.666... and 10.

I3 Divide by any number
(a) between 10 and 20
(b) greater than 20
(c) between 1 and 2
(d) less than 1
(e) greater than 40
(f) less than 0.2

I4 The input must be

(a) greater than 25

(b) between 12.5 and 25

(c) between 2.5 and 5

(d) 2.5

(e) less than 2.5

I5

Input	÷ 20	÷ 2	÷ 0.2
100	5	50	500
20	1	10	100
16	0.8	8	80
8	0.4	4	40

I6

Input	÷ 30	÷ 3	÷ 0.3
60	2	20	200
90	3	30	300
120	4	40	400
15	0.5	5	50

I7

Input	÷ 1	÷ 0.5	÷ 0.2	÷ 0.1
10	10	20	50	100
20	20	40	100	200
2	2	4	10	20
40	40	80	200	400

I8 (a) 5 (20 ÷ 4) (b) 0.2 (4 ÷ 20)

(c) 10 ÷ 5 = 20 ÷ 10   5 ÷ 10 = 10 ÷ 20

(d) 10 ÷ 4

I9 (a) 5 (40 ÷ 8) (b) 0.2 (8 ÷ 40)

(c) 16 ÷ 8 = 40 ÷ 20   8 ÷ 16 = 20 ÷ 40
   20 ÷ 8 = 40 ÷ 16   8 ÷ 20 = 16 ÷ 40

(d) 20 ÷ 8 or 40 ÷ 16

I10 (a) 10 (50 ÷ 5) (b) 0.1 (5 ÷ 50)

(c) None (d) 50 ÷ 20

I11 (a) 3 (24 ÷ 8) (b) 0.3333... (8 ÷ 24)

(c) 16 ÷ 8 = 24 ÷ 12   8 ÷ 16 = 12 ÷ 24
   12 ÷ 8 = 24 ÷ 16   8 ÷ 12 = 16 ÷ 24

(d) 20 ÷ 8 or 40 ÷ 16

I12 (a) 6 (42 ÷ 7) (b) 0.1666... (7 ÷ 42)

(c) None (d) 42 ÷ 12

I13 (a) 40 (16 ÷ 0.4) (b) 0.025 (0.4 ÷ 16)

(c) None (d) 2 ÷ 0.8

I14 (a) 8 (4 ÷ 0.5) (b) 0.125 (0.5 ÷ 4)

(c) 4 ÷ 2 = 2 ÷ 1 = 1 ÷ 0.5
   2 ÷ 4 = 1 ÷ 2 = 0.5 ÷ 1
   4 ÷ 1 = 2 ÷ 0.5
   1 ÷ 4 = 0.5 ÷ 2

(d) 4 ÷ 1 and 2 ÷ 0.5 both give 4;
   2 ÷ 1 and 1 ÷ 0.5 both give 2.

I15 (a) 50 (2 ÷ 0.04) (b) 0.02 (0.04 ÷ 2)

(c) 1 ÷ 2 = 0.5 ÷ 1   2 ÷ 1 = 1 ÷ 0.5

(d) 2 ÷ 0.5 gives 4;
   2 ÷ 1 and 1 ÷ 0.5 both give 2.

I16 (a) 1000 (50 ÷ 0.05)

(b) 0.001 (0.05 ÷ 50)

(c) None (d) 0.2 ÷ 0.05

I17 (a) 500 (30 ÷ 0.06)

(b) 0.002 (0.06 ÷ 30)

(c) None (d) 1.2 ÷ 0.3

J Mixed questions (p 116)

J1 £9.01 ÷ 3.4 = £2.65

J2 The box of 12 is the better deal, because 1 popper costs 160 ÷ 12 = 13.3p (to 1 decimal place), compared with 13.5p in the box of 10.

J3 0.15 × £6.80 = £1.02

J4 £1.30 (1 kg of biscuits costs £0.88 ÷ 0.4 = £2.20.)

J5 10 tins in store A cost £3.60.
10 tins in B cost £3.70.
So A gives the better deal.

What progress have you made? (p 117)

1 (a) 7.35 (b) 447 (c) 23.6
(d) 0.0306

2 (a) 22.8 (b) 9 (c) 2.12
(d) 1.92

3 (a) 4.7 (b) 0.80

4 (a) 2.25   (b) 4.4125

  (c) 2.99   (d) 0.7675

5 (a) Shop A (62p per litre) gives more for your money than Shop B (63.3… p per litre).

  (b) 10p

6 (a) Greater (b) Greater (c) Less

  (d) Less (e) Greater (f) Less

Practice booklet

Sections A and B (p 34)

1 $68 \div 10 = 6.8$

  $68 \times 10 = 680$ $680 \div 10 = 68$

  $6.8 \times 100 = 680$ $680 \div 100 = 6.8$

  $0.57 \times 100 = 57$ $57 \div 100 = 0.57$

2 (a) 6.5   (b) 0.0183   (c) 0.0213

  (d) 54   (e) 4.6   (f) 0.009 46

3 (a) 15.6   (b) 0.96

  (c) 112   (d) 2.8

4 $0.08 \times 300 = 24$ $0.08 \times 500 = 40$

  $0.5 \times 40 = 20$ $0.6 \times 40 = 24$

  $0.6 \times 500 = 300$

Sections C and D (p 34)

1 (a) 85.35   (b) 7.11   (c) 26.07

  (d) 1.50   (e) 24.00

2 (a) 44.286   (b) 4.09   (c) 0.020

  (d) 0.70

3 (a) 0.1125   (b) 0.274   (c) 0.65

  (d) 0.467   (e) 0.359   (f) 0.565

  (g) 0.922   (h) 0.0625

Sections E and F (p 35)

Each answer should be preceded by the pupil's estimate.

1 $88.6 - 69.3 = 19.3\,g$

2 $280 \times 9.54 = 2671.20$ rand

3 $125 \times 0.176 = 22$ dollars

4 Total weight $= 29.2 + 17.7 = 46.9\,kg$
When scales balance, on each side there is $46.9 \div 2 = 23.45\,kg$
So amount transferred $= 29.2 - 23.45 = 5.75\,kg$
(Alternatively, transfer half of the difference between 29.2 and 17.7)

5 (a) (i) $11.85 \times 11.56 + 11.85 \times 8.24 = £234.63$

    (ii) $8.95 \times 11.56 + 8.95 \times 8.24 = £177.21$

    (iii) $11.85 \times 11.56 + 8.95 \times 8.24 = £210.73$

    (iv) $8.95 \times 11.56 + 11.85 \times 8.24 = £201.11$

  (b) $234.63 - 177.21 = £57.42$

  (c) $210.73 - 210.11 = £9.62$

Section G (p 36)

1 (a) £0.77 per litre (b) £2.56 per kg

  (c) £0.96 per litre (d) £3.20 per kg

  (e) £4.60 per kg (f) £0.98 per litre

2 (a) $£1.62 \div 1.5 = £1.08$ per kg
    $£2.20 \div 2 = £1.10$ per kg
    The first pack gives more for your money.

  (b) $£1.85 \div 1.6 = £1.156\,25$ per litre
    $£4 \div 3.5 = £1.142\,85...$ per litre
    The second pack gives more for your money.

3 (a) Clacton: £0.2714... per mile
    Southend: £0.2425 per mile

  (b) London to Southend has the lower cost per mile.

Section H (p 37)

1 (a) 18.30 zloty (b) 152.50 zloty

  (c) 277.55 zloty (d) 4.27 zloty

2 (a) $5 \times 0.80 = £4$

  (b) $38.80 \times 0.80 = £31.04$

  (c) $0.65 \times 0.80 = £0.52$

3 Kate spends $(35 \times £0.88) + (25 \times £0.36)$
  $= £39.80$.

  She receives $60 \times £0.90 = £54$.

  Her profit is $£54 - £39.80 = £14.20$.

4 Cost of 1 g in first jar =
  $£1.50 \div 500 = £0.003$

  Cost of 1 g in second jar =
  $£2.50 \div 800 = £0.003125$

  The first jar is better value for money.

  (You could also do it by finding cost of
  100 g in each jar, and other methods.)

5 $126 \div 2.20 = £57.27$ (to the nearest
  penny)

6 (a) $4.65 \times £3.49 = £16.23$ (to the nearest
    penny)

  (b) $£13 \div 4.65 = £2.80$ per metre (to the
    nearest penny)

  (c) 3.50 m at the normal price costs
    $3.50 \times £3.49 = £12.22$.
    This is cheaper than buying the
    remnant.

Section I (p 38)

1

Input	÷ 40	÷ 4	÷ 0.4
800	**20**	200	**2000**
4	**0.1**	1	**10**
20	**0.5**	5	**50**
36	**0.9**	9	**90**

2

Input	÷ 5	÷ 50	÷ 0.5
100	**20**	2	200
30	**6**	0.6	60
50	**10**	1	**100**
2000	**400**	40	**4000**

3 (a) Bigger   (b) Smaller
  (c) Bigger   (d) Smaller
  (e) Bigger   (f) Smaller
  (g) Smaller   (h) Bigger

4 (a) 7   (b) 12   (c) 80
  (d) 0.8   (e) 400   (f) 47

5 (a) $6 \div 0.03$ (b) $0.03 \div 6$

Section J (p 39)

1 (a) £19.20   (b) £1.35

  (c) With the 18 month subscription,
    each magazine costs
    $£25 \div 18 = £1.39$ (to the nearest
    penny).
    This is more expensive than the 12
    month subscription.

2 (a) Garage A charges $£26.29 \div 30.6$
    $= £0.8591\ldots$ per litre.
    Garage B charges $£42.28 \div 49.8$
    $= £0.8499\ldots$ per litre.
    So B's petrol is cheaper.

  (b) 20 litres would cost $20 \times £0.8499\ldots$
    $= £16.98$ to the nearest penny.

3 Liverpool: 47.7p per mile
  Manchester: 53.5p per mile
  Newcastle: 31.6p per mile
  Penzance: 25.2p per mile

4 The total cost is £116.42 so she cannot
  get a reduction of £10.
  The maximum reduction is £9, which she
  can get in several ways, for example
  $32.60 + 22.80 + 9.89 = 65.29$
  (reduction of £5)
  $15.35 + 17.48 + 18.30 = 51.13$
  (reduction of £4)

16 Gravestones

p 118 **A** What gravestones tell us	Reading from a table
p 119 **B** Making a frequency table	Grouping, tallying
p 120 **C** Comparing charts	Ways of showing frequencies
p 122 **D** Charts galore!	Good and bad charts
p 124 **E** Testing a hypothesis	
p 125 **F** Modal group	

Essential
Sheet 93
Practice booklet page 40

A What gravestones tell us (p 118)

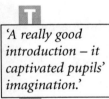

◊ In some circumstances, such as a recent bereavement, sensitive handling is needed.

'A really good introduction – it captivated pupils' imagination.'

The topic obviously lends itself to locally based practical work. The sample you get from a graveyard is only of people rich enough to afford a gravestone.

A3 There may be discussion about which months are in winter.

B Making a frequency table (p 119)

◊ Ask pupils to think of two different ways of tallying from a list.

• Go through the list, tallying all those in the 0–9 group first; then do the 10–19 group, and so on.

• Go through the list once only, tallying into the correct group as you go.

Items are less likely to be missed out if the second way is used.

◊ Pupils may already be familiar with grouping tally marks in fives: ЖЖ

C Comparing charts (p 120)

◊ This can be done in pairs or as a class discussion.

◊ Two ways of drawing a bar chart are shown. In the first, age is treated as discrete and a gap is left between bars. In the second, age is treated as continuous with no gaps between bars, so that a rule is needed for the boundaries.

D Charts galore! (p 122)

Pupils look at good and bad ways of displaying data.

◊ This section will mean more if pupils can use a spreadsheet and draw charts for themselves.

◊ The tilted pie chart can make if difficult to judge the relative sizes of sectors. It loses the empty age groups. It gives no idea of the total sample, but shows the proportions.

The cobweb diagram is hard to read. The shading has no meaning.

The bar chart clearly summarises the data but there is no point in dividing the frequency scale into fifths.

The line graph or frequency polygon retains all the information but can be confusing to interpret. The joining lines have no meaning; they just guide the eye.

The doughnut diagram loses the 10–19 and 20–29 data but shows a gap labelled 30–39!

The 3-D bar chart retains all the information, but the frequency is hard to read.

E Testing a hypothesis (p 124)

Sheet 93

The idea of a hypothesis is a subtle one, but important. You could try describing briefly other contexts and ask for suggested hypotheses that might be tested (for example, 'diet and progress in school', where one hypothesis could be 'children who eat plenty of vegetables do better in school tests'). Emphasise that a hypothesis may turn out to be false.

Challenge (p 124)

John Millington died on 1 September 1694 aged 54.

If he died before his birthday that year then he was born in 1639. If he died after his birthday that year then he was born in 1640.

A What gravestones tell us (p 118)

A1 Probably November 1757. Remember that people are not generally buried on the day they die.

A2 Roughly 90 years

A3 It depends on what are the 'winter months'.

J F M A M J J A S O N D
1   2 1   2 2 1   1 3 1

B Making a frequency table (p 119)

B1

Age (in years)	Tally	Frequency
0–9	ⅢⅡ I	6
10–19		0
20–29		0
30–39		0
40–49	I	1
50–59	III	3
60–69	III	3
70–79	I	1

The table shows that the people tended to die young or old but not in between.

C Comparing charts (p 120)

C1 There are no missing bars. They have zero height.

C2 70–80

C3 There are no people who died in these age groups.

C4 (a) Pie chart
(b) Either bar chart
(c) Either bar chart

E Testing a hypothesis (p 124)

E1 Some pupils will not manage to make a hypothesis of their own. In E2 they can test one of those already suggested.

Possible hypotheses include
'Nobody lived beyond 71.'
'Nobody died aged between 10 and 39.'
'Most people died in November.'
'Boys stood a better chance of surviving childhood.'

E2 'Most deaths occurred in the winter months.'

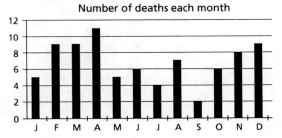

Number of deaths each month

If we define the winter months to be from November to April then the hypothesis seems to be confirmed with 51 dying in that period, 30 outside it.

'Most deaths occurred in the under 10 age group.'

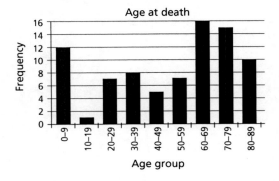

Age at death

This hypothesis is not confirmed. The bar chart shows that most deaths occurred in the 60–69 age group. It is worth commenting that infant mortality was high.

'If you reached 20, there was a good chance of living to 60.'

Of the 68 people who reached age 20, 41 reached age 60. So the hypothesis is confirmed.

F Modal group (p 125)

F1 (a)

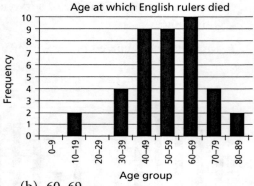

(b) 60–69

(c) The chart shows that the rulers did not die particularly young. (This could be biased because you might have had to wait for an old parent to die before becoming ruler.) Most managed to get into their forties. (It would be interesting to compare this with the general population.)

Practice booklet

Section F (p 40)

1 Pupils may choose groups of a different size from those shown here.

Without fertiliser

Weight of turnips (grams)	Tally	Frequency
100–149	卌 I	6
150–199	卌 卌 II	12
200–249	卌 II	7
250–299	IIII	4
300–349	I	1

With fertiliser

Weight of turnips (grams)	Tally	Frequency
100–149	II	2
150–199	IIII	4
200–249	卌 I	6
250–299	卌 卌	10
300–349	卌 III	8

2

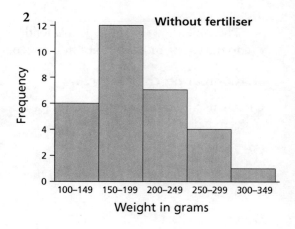

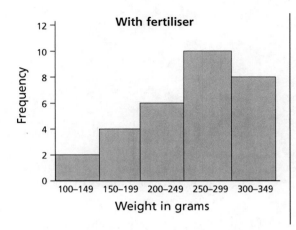

With fertiliser

Frequency / Weight in grams

3 Without fertiliser: 150 g–199 g
  With fertiliser: 250 g–299 g

4 The fertiliser does give bigger turnips.
  Over half the turnips grown without fertiliser are under 200 grams whereas when fertiliser has been used about two-thirds are over 250 grams.

Review 2 (p 126)

1 (a) $u = 4$ (b) $v = 3$ (c) $w = 2\frac{1}{2}$

2 (a) $54 = 2 \times 3 \times 3 \times 3$

(b) 18 (c) 108

3 $w = 2r + 2$

4 (a) 1.25 m

(b) (i) 64p (ii) £1.60

5 (a)

(b)

6 (a) 864 (b) 30 (c) 4

7 (a) 4 (b) 12 (c) 36

8 (a) Foot length = 3.3 × index finger length

(b) 31.5 cm (c) 9.1 cm

9 $a = 143°$, $b = 91°$, $c = 89°$, $d = 123°$, $e = 53°$

10 Largest $5.6 \div 0.4$ Smallest $0.4 \div 5.6$

11 $w = 3r + 2$ (w = number of white tiles, r = number of red tiles)

12 $x = 14°$

13 (a) A 1p coin has a diameter of 2 cm. So in 1.6 km we can fit 80 000 coins, or £800 worth.

(b) A typical heart would beat about 70 times per minute. So a million heart beats would last about 14 000 minutes or about 10 days.

14 (a) There is 1 three by three square, 4 two by two squares and 9 one by one squares, a total of $1^2 + 2^2 + 3^2 = 14$ squares.

(b) There is 1 ten by ten, 4 nine by nine, 9 eight by eight, 16 seven by seven, 25 six by six etc, a total of $1^2 + 2^2 + 3^2 + 4^2 + 5^2 + 6^2 + 7^2 + 8^2 + 9^2 + 10^2 = 385$ squares.

15 There are no unique answers, but …
John may be going up in 2s.
Paul may be adding the previous two numbers together.
Sue may be going 'multiply by 2, add 2' alternately.

16 (a)
```
    9 5 6 7
  + 1 0 8 5
  ---------
  1 0 6 5 2
```
S = 9, E = 5, N = 6, D = 7, M = 1, O = 0, R = 8, Y = 2

(b)
```
    9 6 2 3 3
  + 6 2 5 1 3
  -----------
  1 5 8 7 4 6
```
C = 9, R = 6, O = 2, S = 3, A = 5, D = 1, N = 8, G = 7, E = 4

17 7:10 a.m.

Mixed questions 2 (Practice booklet p 41)

1

	Swimming	Climbing
Male	**7**	**5**
Female	**3**	**6**

2 (a) $n = 9$ (b) $n = 6$

3 (a)

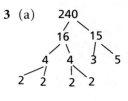

   (b)   243
           /   \
         81   3
          |
          9   9
         /| |\
        3 3   3 3

   (c)   245
          /   \
        35   7
        / \
       7   5

4 $w = 2b - 2$

5 Cost of 1 bar in
      shop A = £2.69 ÷ 3 = 89.7p
      shop B = £3.49 ÷ 4 = 87.25p
      shop C = £4.36 ÷ 5 = 87.2p
The bars in shop C are cheapest.

6 (a) 7

   (b) The 8 animals in the 0–19 age range
        have become 8 in the 10–29 age range,
        so none have died.

   (c) 6

17 Area and perimeter

Most of the work involves rectangles. There is introductory work on right-angled triangles.

p 129 **A** Exploring perimeters

p 130 **B** Rectangles

p 132 **C** Right-angled triangles

p 134 **D** Using decimals

p 135 **E** Using more decimals

p 136 **F** Tenths and hundreds

Essential

Centimetre squared paper

Practice booklet pages 43 to 46

A Exploring perimeters (p 129)

◊ These investigations are in order of increasing difficulty. Pupils can go as far with them as they can manage, but all should become familiar with the word *perimeter* and should come to see that a particular number of squares (producing shapes of a fixed area) can give rise to different perimeters. The perimeters are all even numbers of centimetres (investigation 3) because before the squares are put together their total perimeter is an even number of centimetres (actually a multiple of 4) and each time two edges are put together the total perimeter falls by 2 centimetres, remaining even.

The chart on the next page shows the possible perimeters for investigations 2, 4 and 5.

The maximum perimeter for a given number of squares n is $2n + 2$ or $2(n + 1)$ and arises when all the squares are in a straight line or form a shape 'one square wide' with bends in it (the least 'compact' arrangement). Some pupils should be able to explain, in their own words, why such a relationship applies. This linear relationship is shown by the upper right-hand edge of the block of ticks on the chart.

Perimeter

Number of squares	4	6	8	10	12	14	16	18	20	22	24	26	28	30	32	34	36
1	✓																
2		✓															
3			✓														
4			✓	✓													
5				✓	✓												
6				✓	✓	✓											
7						✓	✓	✓									
8						✓	✓	✓	✓								
9						✓	✓	✓	✓	✓							
10							✓	✓	✓	✓	✓						
11							✓	✓	✓	✓	✓	✓					
12							✓	✓	✓	✓	✓	✓	✓				
13								✓	✓	✓	✓	✓	✓	✓			
14								✓	✓	✓	✓	✓	✓	✓	✓		
15								✓	✓	✓	✓	✓	✓	✓	✓	✓	
16								✓	✓	✓	✓	✓	✓	✓	✓	✓	✓
17									✓	✓	✓	✓	✓	✓	✓	✓	✓

The minimum perimeter is shown by the left-hand edge of the block of ticks. This relationship is not linear but pupils may be able to give their own explanations of why the steps shown with arrows in the chart occur.

B Rectangles (p 130)

C Right-angled triangles (p 132)

C5 The subtraction method will be found easier.

D Using decimals (p 134)

E Using more decimals (p 135)

F Tenths and hundredths (p 136)

T   ◊ Multiplying decimals with two or more digits appears again in the next decimals unit. Here area is used to help explain the results obtained.

B Rectangles (p 130)

B1 (a) The pupil's decision
   (b) The area of the red rectangle $(20\,\text{cm}^2)$ is less than half that of the blue rectangle $(42\,\text{cm}^2)$.

B2 (a) $48\,\text{cm}^2$ (b) $50\,\text{cm}^2$ (c) $73\,\text{cm}^2$

B3 $48\,\text{cm}^2$

B4 $15\,\text{cm}$

B5 $66\,\text{cm}$

B6 $94\,\text{cm}$

B7 (a) $232\,\text{cm}^2$ (b) $372\,\text{cm}^2$

B8 (a) $100\,\text{cm}$ by $100\,\text{cm}$
   (b) $10\,000\,\text{cm}^2$

B9 $664\,\text{m}^2$

C Right-angled triangles (p 132)

C1 (a) $\frac{1}{2}$ (b) $3\,\text{cm}^2$

C2 (a) $20\,\text{cm}^2$ (b) $24\,\text{cm}^2$ (c) $21\,\text{cm}^2$
   (d) $20\,\text{cm}^2$ (e) $28\,\text{cm}^2$

C3 (a) $55\,\text{cm}^2$ (b) $77\,\text{cm}^2$ (c) $49.5\,\text{cm}^2$
   (d) $61.5\,\text{cm}^2$ (e) $102\,\text{cm}^2$

C4 (a) $10.5\,\text{cm}^2$ (b) $10.5\,\text{cm}^2$

C5 (a) $10\,\text{cm}^2$ (b) $12\,\text{cm}^2$
   (c) $8\,\text{cm}^2$ (d) $16\,\text{cm}^2$

C6 (a) $63\,\text{cm}^2$ (b) $139.5\,\text{cm}^2$

D Using decimals (p 134)

D1 (a) $13.5\,\text{cm}^2$ (b) Check

D2 (a) $13\,\text{cm}^2$ (b) $16.5\,\text{cm}^2$ (c) $30\,\text{cm}^2$
   (d) $36\,\text{cm}^2$ (e) $10.8\,\text{cm}^2$ (f) $24.3\,\text{cm}^2$

D3 $9.35\,\text{m}$

D4 (a) $4.7\,\text{cm}$ (b) $5.8\,\text{cm}$
   (c) $6.5\,\text{cm}$ (d) $16.75\,\text{cm}$

D5 (a) $3\,\text{cm}$ (b) $19.8\,\text{cm}^2$

D6 (a) $5.8\,\text{cm}$ (b) $17.6\,\text{cm}$

D7 $4.3\,\text{cm}^2$

D8 $4.4\,\text{cm}^2$

D9 (a) $\frac{1}{4}$ (b) $0.25\,\text{cm}^2$

E Using more decimals (p 135)

E1 $29.25\,\text{cm}^2$

E2 (a) $5.25\,\text{cm}^2$ (b) $21.25\,\text{cm}^2$
   (c) $33.75\,\text{cm}^2$

E3 (a) $41.5\,\text{cm}^2$ (b) $26.125\,\text{cm}^2$
   (c) $57.375\,\text{cm}^2$

F Tenths and hundredths (p 136)

F1 (a) 24 (b) 0.24
   (c) $0.6 \times 0.4 = 0.24$

F2 (a) $0.35\,\text{m}^2$ (b) $0.72\,\text{m}^2$
   (c) $0.4(0)\,\text{m}^2$

F3 (a) $1.8\,\text{cm}^2$ (b) $0.42\,\text{cm}^2$ (c) $0.1\,\text{cm}^2$

F4 She is wrong. $0.2 \times 0.4 = 0.08$
   (A doormat $0.2\,\text{m}$ by $0.4\,\text{m}$ would cover $\frac{8}{100}$ of a square metre.)

F5 (a) $0.06\,\text{cm}^2$ (b) $0.05\,\text{cm}^2$ (c) $0.12\,\text{cm}^2$
   (d) $0.2\,\text{cm}^2$ (e) $0.4\,\text{cm}^2$ (f) $0.09\,\text{cm}^2$

F6 $4.86\,\text{cm}$

F7 (a) $11.48\,\text{cm}^2$ (b) $8\,\text{cm}^2$
   (c) $16.2\,\text{cm}^2$ (d) $21.58\,\text{cm}^2$
   (e) $48.88\,\text{cm}^2$ (f) $46.24\,\text{cm}^2$

F8 Measurements may differ slightly because of printing distortion. Lengths are in centimetres.

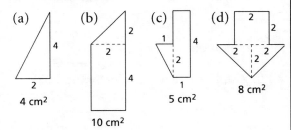

(a) 4 cm² (b) 10 cm² (c) 5 cm² (d) 8 cm²

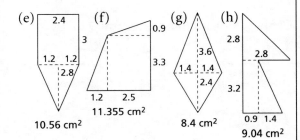

(e) 10.56 cm² (f) 11.355 cm² (g) 8.4 cm² (h) 9.04 cm²

What progress have you made? (p 128)

1 (i) 41 cm²   (ii) 32 cm

2 24.5 cm

3 13.5 cm²

4 (a) 48.75 cm²   (b) 11.04 cm²
(c) 0.08 m²

5 69.26 cm²

Practice booklet

Section B (p 43)

1 (a) 125 cm² (b) 88 cm² (c) 156 cm²

2 (a) 376 cm² (b) 382 cm² (c) 380 cm²

3 500

4 0.8 cm

5 (a) 45 000 cm²   (b) 45 m

Section C (p 44)

1 (a) 7.5 cm² (b) 14 cm² (c) 24 cm²

2 (a) 210 m² (b) 72 m² (c) 252 m²

3 (a) 5.5 cm² (b) 16 cm² (c) 14.5 cm²

Section D (p 45)

1 (a) 16.8 cm² (b) 11.8 cm² (c) 13.5 cm²

2 (a) 18 m² (b) 1.8 m² (c) 0.18 m²
(d) 1.12 m² (e) 2.52 m² (f) 12 m²

3 (a) 2.9 cm   (b) 13.8 cm

4 (a) 3 cm   (b) 14. 4 cm²

Section E (p 45)

1 (a) 6.25 cm²   (b) 16.5 cm²
(c) 22.75 cm²   (d) 9 cm²
(e) 11.25 cm²   (f) 11.25 cm²

2 (a) 48.75 cm²   (b) 60.125 cm²

Section F (p 46)

1 (a) 0.35 m² (b) 0.36 m² (c) 0.4 m²

2 3.78 m²

3 (a) 2.4 m², 6.2 m
(b) 1.68 m², 6.2 m
(c) 2.7 m², 6.6 m
(d) 0.72 m², 3.4 m
(e) 51.84 m², 28.8 m
(f) 3.1 m², 13.4 m

4

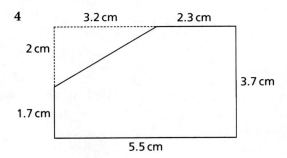

Area 17.15 cm²

Negative numbers 1

7S/14, 7S/48

	p 139 **A** Temperature	
T	p 140 **B** Changes	Simple additions and subtractions
T	p 142 **C** The talent contest	Adding and subtracting a negative number

	Optional
	Large cards with numbers from ⁻9 to 9
Practice booklet pages 47 to 49	

A Temperature (p 139)

B Changes (p 140)

◊ Ask pupils how they see, for example, 2 – 5 on the number line. Help them to associate additions and subtractions with moves on the number line: 2 – 5 as 'start at 2, go down 5'.

◊ It is a good idea to distinguish between the negative sign ⁻ and the subtract sign –, at least to start with.

C The talent contest (p 142)

This introduces adding and subtracting a negative number.

Optional: Large cards with numbers from ⁻9 to 9

◊ The work is more memorable if you have a real contest!

◊ Focus first on adding together a set of numbers which includes a negative number (which pulls the total score down). Check that pupils realise that, for example, ⁻3 + 5 is equal to 5 + ⁻3.

◊ When addition is understood, get pupils to think about what happens to a total score when a negative number is removed (subtracted), as in the lower pictures on page 143.

'This section went well as we did it practically and had joke-telling competitions and all scores were combinations of negative and positives.'

Ⓐ Temperature (p 139)

A1 (a) $^-3°C$ (b) $2°C$ (c) $^-3°C$ (d) $^-15°C$

A2 (a) 10 degrees (b) June, July, August
(c) November, December, January

A3 (a) (i) Helium (ii) Radon
(b) Argon
(c) Nitrogen (than argon)
Helium (than neon)

Ⓑ Changes (p 140)

B1 (a) $^-2 + 3 = \mathbf{1}$ (b) $^-6 + 2 = ^-\mathbf{4}$
(c) $^-8 + 8 = \mathbf{0}$ (d) $^-8 + 10 = \mathbf{2}$
(e) $^-4 + 7 = \mathbf{3}$ (f) $^-1 + 6 = \mathbf{5}$

B2 (a) $3 - 2 = \mathbf{1}$ (b) $3 - 4 = ^-\mathbf{1}$
(c) $^-5 - 4 = ^-\mathbf{9}$ (d) $^-3 - 3 = ^-\mathbf{6}$
(e) $0 - 4 = ^-\mathbf{4}$ (f) $4 - 7 = ^-\mathbf{3}$
(g) $^-7 - 4 = ^-\mathbf{11}$ (h) $^-3 - 0 = ^-\mathbf{3}$
(i) $^-5 - 2 = ^-\mathbf{7}$

B3 (a) $^-3 + 2 = ^-\mathbf{1}$ (b) $^-2 - 8 = ^-\mathbf{10}$
(c) $^-5 + 4 = ^-\mathbf{1}$ (d) $^-6 - 2 = ^-\mathbf{8}$
(e) $3 - 7 = ^-\mathbf{4}$ (f) $^-7 + 3 = ^-\mathbf{4}$

B4 (a) $2 - 6 = ^-4$ (b) $1 - 4 = ^-3$
(c) $^-2 + 3 = 1$ (d) $0 - 3 = ^-3$

B5 (a) $2 - \mathbf{3} = ^-1$ (b) $2 - \mathbf{4} = ^-2$
(c) $2 - \mathbf{5} = ^-3$ (d) $\mathbf{2} - 7 = ^-5$
(e) $^-4 - \mathbf{1} = ^-5$ (f) $\mathbf{7} - 5 = 2$
(g) $\mathbf{8} - 10 = ^-2$ (h) $\mathbf{8} - 4 = 4$

B6 (a) $^-\mathbf{8} + 3 = ^-5$ (b) $^-\mathbf{2} + 4 = 2$
(c) $4 + \mathbf{3} = 7$ (d) $4 - \mathbf{11} = ^-7$

B7 (a) $50 - 60 = ^-\mathbf{10}$
(b) $^-30 - 50 = ^-\mathbf{80}$
(c) $^-\mathbf{40} - 60 = ^-100$
(d) $^-90 + \mathbf{80} = ^-10$
(e) $15 - \mathbf{115} = ^-100$
(f) $100 - \mathbf{121} = ^-21$
(g) $\mathbf{60} - 18 = 42$
(h) $^-63 + \mathbf{48} = ^-15$

B8 (a) $10, 8, 6, 4, \mathbf{2}, \mathbf{0}, ^-\mathbf{2}$ (subtract 2)
(b) $7, 5, 3, 1, ^-\mathbf{1}, ^-\mathbf{3}, ^-\mathbf{5}$ (subtract 2)
(c) $40, 30, 20, 10, \mathbf{0}, ^-\mathbf{10}, ^-\mathbf{20}$
(subtract 10)
(d) $^-2, ^-4, ^-6, ^-8, ^-\mathbf{10}, ^-\mathbf{12}, ^-\mathbf{14}$
(subtract 2)
(e) $^-12, ^-9, ^-6, ^-3, \mathbf{0}, \mathbf{3}, \mathbf{6}$ (add 3)

B9 Five calculations can be made.
$^-4 + 6 = 2$ $^-4 + 5 = 1$ $^-4 + 3 = ^-1$
$^-4 + 2 = ^-2$ $^-4 + 1 = ^-3$

B10 The four calculations which are most likely to be made are
$3 - 6 = ^-3$ $3 - 5 = ^-2$ $3 - 2 = 1$
$3 - 1 = 2$

These are unlikely to be found at this stage: $3 - ^-2 = 5$, $3 - ^-3 = 6$

B11 (a) 5 (b) $^-1$ (c) 6 (d) $^-5$
(e) $^-7$ (f) $^-10$ (g) $^-1$ (h) $^-10$

Ⓒ The talent contest (p 142)

C1 (a) $^-1$ (b) $^-11$ (c) 0 (d) $^-1$

C2 (a) 2 (b) $^-3$ (c) 5 (d) 5
(e) $^-5$ (f) 0 (g) 2 (h) 4
(i) 20

C3 (a) 4 (b) 5 (c) $4 - ^-1 = \mathbf{5}$

C4 (a) $3 - ^-6 = 9$ (b) $3 - ^-3 = 6$

C5 (a) Both have the same effect.
(b) Add 2 to 5.
(c) Add the corresponding positive number.
(d) The pupil's examples

C6 (a) 10 (b) 22 (c) 80
(d) 402 (e) 4.2 (f) 8.22

C7 (a) 5 (b) 17 (c) $^-21$
(d) $^-38$ (e) 206 (f) $^-1.8$

C8 (a) 10 (b) $^-5$ (c) 30
(d) 7.7

C9 (a) 1   (b) 11   (c) $^-2$ (d) 78

C10 The possibilities are too numerous to list.

C11 Many magic squares are possible. The 'magic total' is $^-3$.

C12 The 'magic total' is $^-2$.

What progress have you made? (p 145)

1 (a) $^-3$ (b) $^-3$ (c) $^-14$ (d) $^-80$

2 (a) $2 - 5 = ^-3$ (b) $10 - 15 = ^-5$
    (c) $^-6 - 1 = ^-7$

3 (a) 10   (b) 18   (c) $^-5$ (d) 5

Practice booklet

Section B (p 47)

1 (a) $^-3 + 4 = 1$ (b) $1 - 4 = ^-3$
    (c) $^-4 + 6 = 2$

2 (a) $3 - 6 = ^-3$ (b) $^-2 - 8 = ^-10$
    (c) $0 - 7 = ^-7$ (d) $2 - 4 = ^-2$
    (e) $3 - 5 = ^-2$ (f) $5 - 6 = ^-1$
    (g) $^-7 - 3 = ^-10$ (h) $1 - 2 = ^-1$

3 (a) $^-2 + 10 = 8$ (b) $^-15 + 5 = ^-10$
    (c) $^-1 + 1 = 0$ (d) $^-3 + 11 = 8$

4 (a) $100 - 110 = ^-10$
    (b) $110 - 100 = 10$
    (c) $100 - 300 = ^-200$
    (d) $0 - 10 = ^-10$
    (e) $25 - 100 = ^-75$
    (f) $150 - 250 = ^-100$
    (g) $^-20 - 13 = ^-33$
    (h) $^-1 - 8 = ^-9$
    (i) $^-13 + 7 = ^-6$

5 (a) $22 - 150 = ^-128$
    (b) $^-101 + 67 = ^-34$
    (c) $^-11 + 19 = 8$

Section C (p 48)

1 (a) 2   (b) 3   (c) 1   (d) $^-1$

2 (a) 1   (b) 0   (c) $^-7$ (d) 0
    (e) $^-1$ (f) 3

3 (a) $^-1 + ^-3 = ^-4$ (b) $6 + ^-2 = 4$
    (c) $4 + ^-3 + ^-1 = 0$ (d) $^-4 + 4 = 0$
    (e) $5 + ^-6 + 2 = 1$ (f) $^-1 + ^-1 + 0 = ^-2$

4 (a) 1   (b) $^-4$ (c) $^-4$
    (d) $^-2$ (e) 1   (f) $^-11$

5 (a) 8   (b) 12   (c) $8 - ^-4 = 12$

6 (a) 14   (b) $10 - ^-4 = 14$

7 (a) $12 - ^-3 = 15$ (b) $6 - ^-5 = 11$
    (c) $^-1 - ^-6 = 5$

8 (a) 15   (b) 18   (c) 4   (d) 1
    (e) 0   (f) 5

9 (a) $^-5$ (b) 1   (c) $^-3$ (d) 3
    (e) 3   (f) $^-1$ (g) $^-3$ (h) $^-2$
    (i) $^-14$ (j) $^-8$ (k) 135   (l) $^-57$

10 (a) $^-5 - ^-17 = 12$
    (b) $2 - ^-8 = 10$
    (c) $8 - ^-12 = 20$
    (d) $^-16 - ^-4 = ^-12$
    (e) $^-2 + ^-10 = ^-12$
    (f) $^-2 - 10 = ^-12$
    (g) $0 + ^-89 = ^-89$
    (h) $^-15 - ^-30 = 15$
    (i) $^-5 + ^-45 = ^-50$
    (j) $^-70 - ^-170 = 100$
    (k) $^-68 - ^-23 = ^-45$
    (l) $0 - ^-45 = 45$

Spot the rule

The teacher-led, whole-class activity that starts section A is the most important activity in this unit.

At first sight, it may seem perverse not to use input numbers in order, increasing by 1. But this encourages just one method for solving these rule puzzles. An equally effective method may be to ask 'What happens to 100? to 1000?', and pupils may use others.

p 146 **A** Finding rules	Finding a rule that connects pairs of numbers	
p 148 **B** Further rules	Using shorthand such as n^2 and $2n^2$	

> **Practice booklet** pages 50 and 51

A **Finding rules** (p 146)

◊ This section starts with a teacher-led activity with the whole class. This is especially engaging and effective if carried out in complete silence, as the following account from a school makes clear.

'It was the last lesson on Friday afternoon. I allowed time for settling down and said we were going to play a game … everyone had to be silent, including me!

Without saying anything, I wrote on the board …

2	→	9
5	→	12
3	→	

… and (still without saying anything) invited someone to come and fill in the gap.

3	→	10	☺

As they were right I drew a smiley face.

I then added more numbers each time on the left-hand side. Different pupils came and wrote up numbers on the right-hand side.

3	→	10	☺		
7	→	15	☹	14	☺
10	→	17	☺		
0	→	7	☺		
25	→				

If the pupil's number was wrong I drew a sad face and signalled another pupil to try. They liked big numbers, so I continued with larger ones.

After a while I wrote up …

    number   →

… and received the responses

    number   →   add 7   ☹

                       →   number add on 7   ☺

I was looking for the rule here (which could be in words).

After several numbers had been written on the board and pupils had come up and answered in turn, I wrote up a selection of numbers in the left-hand column which included some easy numbers and some more difficult ones.

Pupils came out and filled	25	→		
in the ones they felt	109	→		
comfortable with.	4	→	11	☺
Decimals and negative	964	→		
numbers were included	0.5	→	7.5	☺
when appropriate.	⁻3.4	→		

After a few games, instead of *number* I introduced the class to using n or another letter. I think it is important to emphasise the use of n for number when pupils are expressing their rules. Initially $n \rightarrow\ + 3$ is quite common instead of $n \rightarrow n + 3$.

Once I had played three or four games with the class I asked the pupils to think up some rules of their own. They then took turns to try out their games with the class.

I found there was a beautiful atmosphere with the lesson progressing with no one ever speaking. This can happen quite naturally and seems to strengthen pupils' focus on puzzling out the rule. It allows time for everyone to think and not just the quick ones.'

◊ After playing the game with the whole class, you could continue in groups.

A variation is to set up a simple spreadsheet to produce outputs automatically – nothing complicated, just a formula such as = A1*2 + 1 hidden in cell B1. Pupils can then just type numbers into A1 and see each corresponding result in B1. You could use graphic calculators in a similar way.

Ⓑ Further rules (p 148)

> 'Pupils found this hard – worthwhile keeping it in, though.'

Ⓐ Finding rules (p 146)

A1 (a)

$n \to n + 11$	
3 →	14
10 →	**21**
30 →	**41**
0 →	**11**
56 →	**67**
32 →	43

(b)

$n \to n \div 5$	
15 →	3
10 →	**2**
20 →	**4**
0 →	**0**
50 →	10

(c)

$n \to 2n + 1$	
3 →	7
6 →	**13**
10 →	**21**
8 →	**17**
5 →	11

(d)

$n \to 10 - n$	
3 →	7
8 →	2
4 →	**6**
9 →	**1**
5 →	**5**
7 →	3

A2 (a) $n \to n - 5$ (b) $n \to n \div 3$
(c) $n \to 9 - n$

A3 (a) It could be any of the rules.
(b) $n \to 4n - 1$

A4 (a) $n \to 5n + 1$
(b) 21 (c) 101 (d) 126 (e) 1

A5

$n \to 4n - 2$	
3 →	10
8 →	**30**
4 →	**14**
9 →	**34**
5 →	18

A6 (a) $n \to 3n + 2$ (b) $n \to 6n - 1$
(c) $n \to 2n + 3$

A7

$n \to \dfrac{n}{4}$	
20 →	5
12 →	**3**
36 →	**9**
100 →	**25**
32 →	8

A8 (a) 2 (b) 5 (c) 2.5
(d) 0.5 (e) 0

A9 (a) $n \to \dfrac{n}{2}$ (b) $n \to 30 - 2n$
(c) $n \to \dfrac{n}{6}$ (d) $n \to 8 - 2n$
(e) $n \to 100 - 4n$ (f) $n \to \dfrac{n}{2}$

Ⓑ Further rules (p 148)

B1

$n \to n^2$	
3 →	9
10 →	**100**
5 →	**25**
4 →	16

B2 (a)

$n \to n^2 + 1$	
10 →	101
5 →	**26**
4 →	**17**
9 →	82

(b)

$n \to \dfrac{n^2}{2}$	
10 →	50
6 →	**18**
8 →	**32**
5 →	12.5

(c)

$n \to 2n^2$	
10 →	200
4 →	**32**
6 →	**72**
5 →	50

B3 (a) $n \to n^2 + 10$

(b) $n \to \frac{n^2}{4}$ or $\left(\frac{n}{2}\right)^2$

(c) $n \to n^2 - 1$

(d) $n \to n^2 + n$

B4 The pupil's game

What progress have you made? (p 149)

1 (a) $n \to 3n + 1$ (b) $n \to \frac{n}{6}$

2

$n \to n^2 - 3$	
5 →	**22**
10 →	**97**
12 →	**141**

3 $n \to n^2 + 3$

Practice booklet

Section A (p 50)

1 (a)

$n \to n - 7$		
10	→	**3**
20	→	**13**
8	→	**1**
33	→	**26**
17	→	10
7	→	0

(b)

$n \to 3n$		
4	→	12
5	→	**15**
8	→	**24**
13	→	**39**
10	→	30
7	→	21

(c)

$n \to 8 - n$		
3	→	5
6	→	**2**
1	→	**7**
$\frac{1}{2}$	→	$7\frac{1}{2}$
4	→	4
0	→	8

2 (a) $n \to \frac{n}{2} - 1$

(b) 2   (c) 5   (d) 14   (e) $4\frac{1}{2}$

3 (a)

$n \to 10n - 2$		
3	→	**28**
4	→	**38**
7	→	**68**
8	→	**78**
10	→	98

(b)

$n \to \frac{n}{3}$		
15	→	**5**
21	→	**7**
60	→	**20**
36	→	**12**
27	→	9

(c)

$n \to 30 - 2n$		
4	→	**22**
6	→	**18**
9	→	**12**
11	→	**8**
8	→	14

4 (a) $n \to (n \times 2) + 1$ or $n \to 2n + 1$

(b) $n \to (4 \times n) - 1$ or $n \to 4n - 1$

Section B (p 51)

1 (a)

$n \to n^2 + 2$		
3	→	**11**
5	→	**27**
8	→	**66**
10	→	**102**
2	→	6
12	→	146

(b)

$n \to n^2 + n$		
1	→	**2**
5	→	**30**
10	→	**110**
6	→	42
9	→	90
4	→	20

(c)

$n \to n \times (n + 1)$		
1	→	**2**
5	→	**30**
10	→	**110**
6	→	42
9	→	90
4	→	20

2 (a) $n \to 3n^2$ and $n \to n^2 + 2$ could be her rules.

(b) The pupil's three rules

(c) $n \to 3n^2$ is being used.

(d) $2 \to$ **12**, $10 \to$ **300**, $7 \to$ **147**

3 (a) $n \to n^2 - 1$ (b) $n \to 10n^2$

Chance

This unit introduces probability through games of chance. Probabilities are based on equally likely outcomes and the work here includes equivalent fractions.

p 150	**A** Chance or skill?	Deciding whether a game involves chance or skill
p 151	**B** Fair or unfair?	Deciding whether a game of chance is fair
p 152	**C** Probability	The probability scale from 0 to 1
p 153	**D** Equally likely outcomes	Writing a probability as a fraction Finding the probability of an event not happening
p 155	**E** Equivalent fractions	
p 157	**F** Choosing at random	
p 159	**G** Revisiting games of chance	Listing outcomes of, for example, two dice

Essential	**Optional**
Dice, counters Sheets 111 to 115	OHP transparencies of sheets 114 and 115
Practice booklet pages 52 to 54	

A **Chance or skill** (p 150)

Dice and counters, sheets 111 to 113 (game boards)

◊ Before discussing and playing the games, you could get pupils talking about chance, e.g. the National Lottery. People often have peculiar ideas about chance. For example, would they write on a National Lottery ticket the same combination as the one which won last week? If not, why not?

You could ask pupils to think about games they know and to discuss the elements of chance and skill in them.

◊ Before playing the games, ask pupils to think about each game in advance and to try to decide from its rules whether it is a game of pure chance, a game of skill, or a mixture.

Some games of skill give an advantage to the first player. Who goes first is usually decided by a process of chance.

◊ You could split the class into pairs or small groups, with each group playing one of the games and reporting on it.

◊ 'Fours' is a game of skill. 'Line of three' is a mixture of chance and skill. 'Jumping the line' appears to involve skill, because you have to decide which counters to move and it looks as if you can get 'nearer' to winning. But it is a game of pure chance. At any stage there is only one number which will enable the player to win. If any other number comes up, whatever the player does leaves the opponent in essentially the same position.

B Fair or unfair? (p 151)

> Dice, counters, sheets 114 and 115
> Optional: Transparencies of sheets 114 and 115

◊ You can start by playing the 'Three way race' several times as a class, with a track on the board.

◊ When pupils play the game themselves, ask them to record the results and then pool the class's results.

◊ Let pupils consider each other's ways of making the game fairer (if they can think of any!). Do they agree that they would be fairer?

Rat races

'The second race works well as a class if a rat number is assigned to a small group of pupils.'

'We had volunteer rats, a bookie, and the rest were punters as the rats moved across the room.'

The 'First rat race' is straightforward (although there may be some pupils who think that 6 is 'harder' to get than other numbers). In the second race you could ask for suggestions for making it fairer, still using two dice. (For example, the track could be shortened for the 'end' numbers – even so, Rat 1 is never going to win!)

For the 'Second rat race', pupils could list possible outcomes to discover that there are more ways to make 7 than there are to make 3, for example. So some scores are more likely than others.

This is taken up in section G.

C Probability (p 152)

◊ Explain first the meanings of the two endpoints of the scale. Something with probability 0 is often described as 'impossible'. However, there are different ways of being impossible and some of them have nothing to do with probability (for example, it is impossible for a triangle to have four sides). So it is better to say 'never happens'. Something with probability 1 always happens, or is certain to happen.

◊ Go through the events listed in the pupil's book and discuss where they go on the scale. The coin example leads to the other especially important point on the scale, $\frac{1}{2}$. Associate this with 'equally likely to happen or not happen', with fairness, 'even chances', etc.

◊ Keep the approach informal. The important thing is to locate a point on the right side of $\frac{1}{2}$, or close to one of the ends when appropriate (for example, in the case of the National Lottery!).

D Equally likely outcomes (p 153)

◊ A spinner is very useful in connection with probability. It shows fractions in a familiar way.

D4 If the pupil's answer for (d) is $\frac{1}{3}$, then they have ignored the inequality of the parts.

Odds

Although some teachers would like to outlaw 'odds', this language is used a lot in the real world. So it may be better to explain the connection, and the difference, between probability and odds.

Bookmakers' odds make an allowance for profit and are not linked to probability in the simple way shown on the pupil's page. It is only 'fair odds' that are so linked.

E Equivalent fractions (p 155)

◊ 'Pie' diagrams can be used to explain why the numerator and denominator are both multiplied by the same number. For example, in the case of $\frac{3}{4}$, each of the quarters can be subdivided into, say, 5 equal parts, giving $\frac{15}{20}$ as an equivalent fraction.

◊ You may need to emphasise that equivalence works both ways: $\frac{3}{6}$ is equivalent to $\frac{1}{2}$ and vice versa.

◊ Some pupils may have a tendency to produce a list of equivalent fractions by doubling the numerator and denominator each time, for example

$$\frac{1}{3} = \frac{2}{6} = \frac{4}{12} = \frac{8}{24} = \cdots$$

Emphasise that this strategy leads to missed fractions, for example $\frac{3}{9}$.

F Choosing at random (p 157)

◊ In some cultures, raffles and all forms of gambling are disapproved of. But if there are no objections you could simulate a raffle in class.

◊ There are some misconceptions which are worth bringing into the open. Some people think that a 'special' number, like 1 or 100, is less likely to win than an 'ordinary' number (because there are fewer 'special' than 'ordinary' numbers).

F4 Part (e) assumes knowledge of factors. You may wish to check pupils' knowledge before they try this question.

F6, 7 These questions should lead to discussion. Pupils may not have a strategy for comparing fractions, but may still give valid reasons for their choices. For example, 'B has twice as many reds as A but more than twice as many greens, so it's worse'.

In F7, pupils may say 'Choose D, because it has 3 more greens and only 2 more reds'. The choice is correct but the reasoning is not. Suppose, for example, bag C had 2 green and 1 red and bag D had 5 green and 3 red. The probability of choosing green from C would be $\frac{2}{3}$, and from D $\frac{5}{8}$. $\frac{2}{3}$ is greater than $\frac{5}{8}$, so C would be the better choice.

G Revisiting games of chance (p 159)

Pupils list the outcomes for the throws of two dice, etc. This leads to explanations for some of the findings in section B.

Listing pairs of outcomes is covered in more depth in a later unit ('No chance').

◊ Make sure that both 1, 5 and 5, 1, for example, appear in the list of pairs.

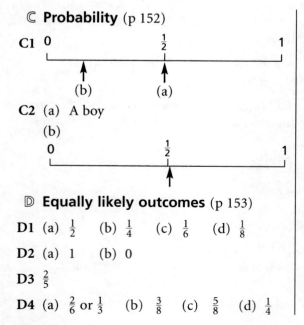

C Probability (p 152)

C1

C2 (a) A boy
(b)

D Equally likely outcomes (p 153)

D1 (a) $\frac{1}{2}$ (b) $\frac{1}{4}$ (c) $\frac{1}{6}$ (d) $\frac{1}{8}$

D2 (a) 1 (b) 0

D3 $\frac{2}{5}$

D4 (a) $\frac{2}{6}$ or $\frac{1}{3}$ (b) $\frac{3}{8}$ (c) $\frac{5}{8}$ (d) $\frac{1}{4}$

D5 (a) $\frac{1}{6}$ (b) $\frac{2}{6}$ or $\frac{1}{3}$ (c) $\frac{3}{6}$ or $\frac{1}{2}$

D6 (a) $\frac{5}{6}$ (b) $\frac{5}{8}$ (c) $\frac{3}{8}$ (d) $\frac{3}{4}$

D7 $\frac{3}{5}$

D8 (a) $\frac{2}{3}$ (b) $\frac{1}{8}$ (c) $\frac{4}{9}$ (d) $\frac{7}{10}$ (e) $\frac{1}{2}$

E Equivalent fractions (p 155)

E1 $\frac{3}{12}$

E2 $\frac{2}{10}$

E3 $\frac{6}{8}, \frac{9}{12}, \frac{12}{16}$

E4 (a) $\frac{1}{2}$ (b) $\frac{3}{4}$ (c) $\frac{1}{4}$ (d) $\frac{3}{4}$ (e) $\frac{3}{4}$

E5 (a) $\frac{1}{4}$ (b) $\frac{3}{8}$ (c) $\frac{2}{3}$ (d) $\frac{3}{7}$ (e) $\frac{1}{5}$

E6 (a) $\frac{2}{3}$ (b) $\frac{5}{8}$ (c) $\frac{2}{5}$
(d) Cannot be simplified (e) $\frac{2}{5}$

E7 (a) $\frac{1}{3}$ (b) Cannot be simplified   (c) $\frac{2}{3}$

(d) Cannot be simplified   (e) $\frac{2}{5}$ (f) $\frac{2}{3}$

(g) $\frac{3}{7}$ (h) Cannot be simplified   (i) $\frac{5}{8}$

(j) $\frac{4}{15}$

E8 $\frac{2}{3}$

F Choosing at random (p 157)

F1 $\frac{1}{50}$

F2 $\frac{4}{200} = \frac{1}{50}$

F3 $\frac{1}{25}$

F4 (a) $\frac{1}{8}$ (b) $\frac{1}{4}$ (c) $\frac{3}{8}$ (d) $\frac{5}{8}$

(e) $\frac{1}{2}$ (f) 0   (g) $\frac{3}{4}$

F5 (a) $\frac{1}{100}$ (b) $\frac{1}{65}$ (c) $\frac{1}{64}$

F6 Sarah should choose bag A.

$\frac{3}{8} = \frac{9}{24}$; $\frac{6}{18} = \frac{1}{3} = \frac{8}{24}$

F7 Dilesh should choose bag D.

$\frac{4}{7} = \frac{48}{84}$; $\frac{7}{12} = \frac{49}{84}$

G Revisiting games of chance (p 159)

G1 (a) 1, 1   1, 2   1, 3   1, 4   1, 5   1, 6
2, 1   2, 2   etc. up to 6, 6

(b) Two even: 9 pairs
Two odd: 9 pairs
One even, one odd: 18 pairs

(c) It explains why C wins most often.

G2 (a)

Total		12	11	10
Number of pairs		1	2	3

9	8	7	6	5	4	3	2
4	5	6	5	4	3	2	1

(b) It explains why the middle numbers win more often.

G3 (a) $\frac{3}{36}$ or $\frac{1}{12}$

(b) 7, with probability $\frac{6}{36}$ or $\frac{1}{6}$

G4 (a) 9   (b) $\frac{9}{36}$ or $\frac{1}{4}$

(c) $\frac{9}{36}$ or $\frac{1}{4}$ (d) $\frac{18}{36}$ or $\frac{1}{2}$

G5 (a) $\frac{4}{36}$ or $\frac{1}{9}$

(b)

Product	1	2	3	4
Probability ($\frac{1}{36}$)	1	2	2	3

5	6	8	9	10	12	15	16	18
2	4	2	1	2	4	2	1	2

20	24	25	30	36
2	2	1	2	1

(c) 0

G6 (a) The game is not fair.
The probability of each difference is:

0	1	2	3	4	5
$\frac{6}{36}$	$\frac{10}{36}$	$\frac{8}{36}$	$\frac{6}{36}$	$\frac{4}{36}$	$\frac{2}{36}$

The probability that Jack wins is $\frac{24}{36}$ and that Jill wins is $\frac{12}{36}$.

(b) One way to modify the game is for Jack to win if the difference is odd.

G7 (a) Sc, Sc   Sc, Pa   Sc, St
Pa, Sc   Pa, Pa   Pa, St
St, Sc   St, Pa   St, St

(b) $\frac{3}{9}$ or $\frac{1}{3}$

G8 All the different orders are equally likely. There are 24 of them.

The probability that Jo will be right is $\frac{1}{24}$.

What progress have you made? (p 161)

1 (a)

(b) It never happens.

(c) It always happens (or it is certain).

(d) See diagram.

2 (a) $\frac{2}{5}$ (b) $\frac{1}{4}$

3 $\frac{4}{80}$ or $\frac{1}{20}$

4 (a) $\frac{3}{5}$ (b) $\frac{5}{9}$ (c) $\frac{2}{5}$

5 (a) 1, 1   1, 2   1, 3   1, 4
2, 1   2, 2   2, 3   2, 4
3, 1   3, 2   3, 3   3, 4

(b) $\frac{3}{12}$ or $\frac{1}{4}$

Practice booklet

Sections D, E and F (p 52)

1 Spinner A

(a) $\frac{4}{8}$ or $\frac{1}{2}$ (b) $\frac{3}{8}$ (c) $\frac{1}{8}$ (d) $\frac{5}{8}$

Spinner B

(a) $\frac{2}{6}$ or $\frac{1}{3}$ (b) $\frac{1}{6}$ (c) $\frac{3}{6}$ or $\frac{1}{2}$ (d) $\frac{5}{6}$

Spinner C

(a) $\frac{1}{3}$ (b) $\frac{1}{3}$ (c) $\frac{1}{3}$ (d) $\frac{2}{3}$

2 (a) A (b) A (c) B

3 Three fractions equivalent to

(a) $\frac{1}{4}$, e.g. $\frac{2}{8}$ $\frac{3}{12}$ $\frac{4}{16}$ $\frac{5}{20}$ $\frac{6}{24}$

(b) $\frac{3}{5}$, e.g. $\frac{6}{10}$ $\frac{9}{15}$ $\frac{12}{20}$ $\frac{15}{25}$ $\frac{18}{30}$

(c) $\frac{5}{8}$, e.g. $\frac{10}{16}$ $\frac{15}{24}$ $\frac{20}{32}$ $\frac{25}{40}$ $\frac{30}{48}$

(d) $\frac{4}{7}$, e.g. $\frac{8}{14}$ $\frac{12}{21}$ $\frac{16}{28}$ $\frac{20}{35}$ $\frac{24}{42}$

4 (a) $\frac{1}{2}$ (b) $\frac{2}{3}$ (c) $\frac{3}{5}$ (d) $\frac{2}{3}$ (e) $\frac{7}{12}$

5 (a) $\frac{25}{55}$ Others are equivalent to $\frac{1}{2}$.

(b) $\frac{9}{15}$ Others are equivalent to $\frac{2}{3}$.

(c) $\frac{35}{50}$ Others are equivalent to $\frac{4}{5}$.

6 (a) $\frac{24}{36} = \frac{2}{3}$ (b) Cannot be simplified

(c) $\frac{30}{140} = \frac{3}{14}$ (d) $\frac{8}{54} = \frac{4}{27}$

7 (a) $\frac{2}{9}$ (b) $\frac{1}{9}$ (c) $\frac{6}{9}$ or $\frac{2}{3}$

(d) $\frac{6}{9}$ or $\frac{2}{3}$ (e) $\frac{3}{9}$ or $\frac{1}{3}$ (f) $\frac{4}{9}$

8 The probability of an orange sweet from A is $\frac{3}{9} = \frac{1}{3} = \frac{10}{30}$.

The probability of an orange sweet from B is $\frac{6}{20} = \frac{3}{10} = \frac{9}{30}$.

So Ann should pick a sweet from A since the probability of picking an orange sweet from that bag is greater.

9 The table shows the probabilities, and the bag from which Rick should pick to have most chance of getting a sweet he likes. Decimal probabilities have been used to help comparison, but equivalent fractions could also be used.

	Blue	Blue or yellow
Bag P	0.3	0.6
Bag Q	0.333	0.5
Bag R	0.25	0.625
Rick's choice	Bag Q	Bag R

Section G (p 54)

1 (a)

(b) 6

(c) $\frac{11}{36}$

2 $\frac{4}{9}$

3 $\frac{4}{9}$

4 $\frac{4}{9}$

*5 If you choose last then you have an advantage because:

if the first player chooses A, you can pick B;

if the first player chooses B, you can pick C;

if the first player chooses C, you can pick A.

In each case the probability you win is $\frac{5}{9}$.

*6 This is a table of all the possible outcomes. The winning number is ringed.

(a) $\frac{8}{27}$ (b) $\frac{11}{27}$ (c) $\frac{8}{27}$

Translation

Essential	Optional
Sheets 123, 124	2 OHP transparencies of sheet 123
Practice booklet page 55	

Ⓐ Describing a translation (p 162)

◊ Emphasise the need to draw the vector from a point on the object to the corresponding point on the image. (Question A2 raises this issue.)

Ⓑ Infinite patterns (p 164)

> Essential: sheets 123, 124
> Optional: 2 OHP transparencies of sheet 123

◊ Some pupils may need to be reminded that the pattern has to be thought of as extending infinitely in all directions. Only then will it map onto itself by a translation.

◊ OHP transparencies are ideal for showing rotation symmetries of an infinite pattern. Use one transparency for the fixed pattern and the other for the tracing.

Ⓐ Describing a translation (p 162)

A1 (a) $\begin{bmatrix} 1 \\ 3 \end{bmatrix}$ (b) $\begin{bmatrix} -1 \\ 5 \end{bmatrix}$ (c) $\begin{bmatrix} 2 \\ -3 \end{bmatrix}$

(d) $\begin{bmatrix} 0 \\ 4 \end{bmatrix}$ (e) $\begin{bmatrix} -1 \\ -4 \end{bmatrix}$ (f) $\begin{bmatrix} 2 \\ 0 \end{bmatrix}$

A2 Donna is wrong. It should be $\begin{bmatrix} 6 \\ 2 \end{bmatrix}$.

A3

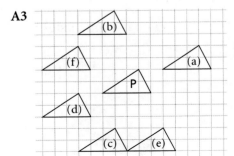

A4 (a) $\begin{bmatrix} 4 \\ 0 \end{bmatrix}$ (b) $\begin{bmatrix} 2 \\ -2 \end{bmatrix}$ (c) $\begin{bmatrix} -2 \\ -2 \end{bmatrix}$

(d) $\begin{bmatrix} -2 \\ 2 \end{bmatrix}$ (e) $\begin{bmatrix} 0 \\ -4 \end{bmatrix}$ (f) $\begin{bmatrix} -2 \\ 3 \end{bmatrix}$

(g) $\begin{bmatrix} -6 \\ -2 \end{bmatrix}$ (h) $\begin{bmatrix} -6 \\ 2 \end{bmatrix}$

A5 (a) $\begin{bmatrix} -4 \\ -3 \end{bmatrix}$

(b) (i) $\begin{bmatrix} -5 \\ 0 \end{bmatrix}$ (ii) $\begin{bmatrix} 1 \\ -4 \end{bmatrix}$

(iii) $\begin{bmatrix} 3 \\ 2 \end{bmatrix}$ (iv) $\begin{bmatrix} -2 \\ 4 \end{bmatrix}$

B Infinite patterns (p 164)

B1 (a) Yes (b) No (c) Yes (d) Yes
(e) Yes (f) No (g) Yes (h) Yes
(i) Yes (j) No (k) Yes (l) Yes

B2 (a) 4; order 4

(b) (i) 2 (ii) 4

(c)
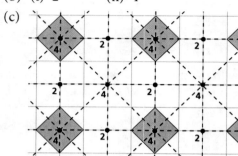

(d) Examples of possible translations:

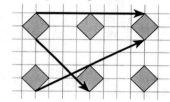

B3

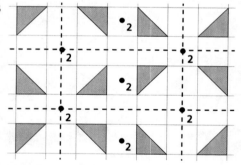

Examples of possible translations:

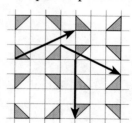

B4

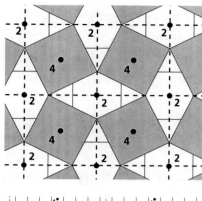

These diagrams show examples of translations:

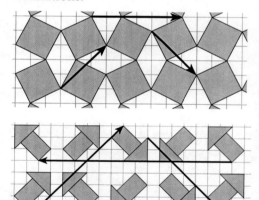

What progress have you made? (p 165)

1 (a) $\begin{bmatrix} 6 \\ -2 \end{bmatrix}$ (b) $\begin{bmatrix} -6 \\ 2 \end{bmatrix}$

Practice booklet

Sections A and B (p 55)

1 (a) $\begin{bmatrix} 5 \\ 1 \end{bmatrix}$ (b) $\begin{bmatrix} 0 \\ 2 \end{bmatrix}$ (c) $\begin{bmatrix} -5 \\ 1 \end{bmatrix}$

   (d) $\begin{bmatrix} -5 \\ -1 \end{bmatrix}$ (e) $\begin{bmatrix} 0 \\ -2 \end{bmatrix}$ (f) $\begin{bmatrix} 5 \\ -1 \end{bmatrix}$

2 (a) (i) No   (ii) No   (iii) Yes
      (iv) Yes   (v) Yes

  (b) (i) No   (ii) No

  (c) (i) 4   (ii) 2   (iii) 4

22 Stretchers

This is a short collection to be selected from: it is not intended that they should all be done together. Some of them are very hard! Pupils should not expect to be able to solve each problem in a short time, as if it was a question in an exercise. They can come back to it later.

Ⓐ Minimal measuring (p 166)

As an extension to question A8, pupils can design a set of weights for when more than two weights may be used.

Ⓑ Fives (p 168)

Ⓒ Tablecloths (p 168)

Ⓐ Minimal measuring (p 166)

A1 (a) The 10 cm and 3 cm strips

(b) 4, 6, 8, 11, 13, 15

(c) (i) 9, 14, 16, 18

(ii) 19

(d) 2, 7, 12, 17

(e) 1, 2, 4, 8, 16
The longest length is 31 cm.

A2 1, 2, 3, 4, 5, 6, 7, 8, 9, 11, 12, 13, 14, 17

A3 1, 3, 9, 27
The longest length is 40 cm.

A4 (a) The pupil's explanation

(b) The pupil's explanation

(c) 1, 2, 3, 4, 5, 6, 8, 10

A5 (a) 1, 2, 3, 4, 5, 7, 8, 9, 10, 11, 12

(b) (i) There are many ways. It is obviously easier if you allow more marks.

(ii) 4 marks. One way is 2, 2, 1, 6, 1.

(iii) Yes, there are 14 ways (including reverses).

A6 (a) 30°, 60°, 90°, 120°, 150°, 180°, 210°, 240°, 270°, 300°, 330°

(b) Four marks are needed.
Two ways are 30°, 90°, 60°, 180° and 30°, 150°, 120°, 60°.

A7 1, 2, 3, 4, 5, 9, 15 kg

A8 1, 2, 3, 4, 5, 15 kg

Ⓑ Fives (p 168)

B1 (a) There are 2 numbers from 1 to 20 that have a five in.

(b) There are 19 numbers from 1 to 100 with a five in.

(c) So the number of numbers from 1 to 1000 with a five in is:

9×19 (19 in each 'century' except 500 to 599)

$+ 100$ (every number from 500 to 599)

$= 271$

B2 From 1 to 100 you write 20 fives (10 fives in the units place and 10 fives in the tens place).

From 1 to 1000 there are
10×20 (all centuries)
$+ 100$ (the extra fives in the hundreds place)
$= 300$.

From 1 to 10 000 there are
$10 \times 300 + 1000 = 4000$
From 1 to 100 000 there are
$10 \times 4000 + 10\ 000 = 50\ 000$
and so on.

© **Tablecloths** (p 168)

C1 The lengths of the sides are both odd, so all the corner squares are red. The rows contain alternately 43 red and 42 white, and 42 red and 43 white. So there are 34 rows each with 43 red and 42 white, and 33 rows each with 42 red and 43 white.

(a) 2848   (b) 2847

C2 The red squares form an array of 35 rows with 46 in each. The white squares form an array of 36 rows with 47 in each row. The number of pink squares can be found by subtracting the red and white from the total.

(a) 1610   (b) 1692   (c) 3301

23 # Number grids

In this unit, pupils solve number grid problems using addition and subtraction. This includes using the idea of an inverse operation ('working backwards').

Algebra arises through investigating number grids. Pupils simplify expressions such as $n + 4 + n - 3$ and produce simple algebraic proofs of general statements. They also simplify expressions such as $2n \times n$.

T	p 169 **A** Square grids	Simple addition and subtraction problems
		Investigating 'diagonal rules' on number grids
	p 170 **B** Grid puzzles	Using the idea of an inverse operation ('working backwards') to solve number puzzles
		Solving more difficult puzzles, possibly by trial and improvement
T	p 171 **C** Algebra on grids	Knowing that, for example, a number 2 more than n is $n + 2$
		Simplifying expressions such as $n + 4 - 3$
	p 173 **D** Grid investigations	Exploring and describing number patterns
		Simple algebraic proofs
T	p 174 **E** Using algebra	Simplifying expressions such as $n + 4 + n - 3$
		Simple algebraic proofs
	p 176 **F** Using multiplication	Simplifying expressions such as $2n \times n$ and $3x \times 4x$

Optional
A4 sheets of paper
Felt-tip pens or crayons

Practice booklet pages 56 to 59

A Square grids (p 169)

The idea of a number grid is introduced. There are many opportunities to discuss mental methods of addition and subtraction.

> Optional: A4 sheets of paper and felt-tip pens or crayons (for 'Human number grids')

Human number grids

This introductory activity does not appear in the pupil's book.

◊ Each pupil or pair of pupils represents a position in a number grid.

Each position will contain a number. (For pupils familiar with spreadsheets, the idea of a 'cell' may help.)

The operations used are restricted to addition and subtraction.

◊ Tables/desks need to be arranged in rows and columns so that the cells form a grid. Explain, with appropriate diagrams, that the class is going to form a human number grid that uses rules to get from a number in one cell to a number in another. A possible diagram is shown below.

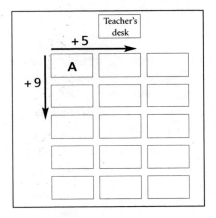

◊ Referring to the numbers is easier if the cells are labelled.
Pupils can discuss how each cell might be labelled, for example:

• A1, A2, B1, … as on a spreadsheet

• A, B, C, …

• or with the pupils' names

◊ Initially, it may help to use only addition or use sufficiently large numbers in cell A to avoid the complication of negative numbers.

◊ Decide on the first pair of rules and ask the pupils in the cell marked A in the diagram to choose a number for that cell.
Discuss how the numbers in other cells are found.
Now ask the pupils in cell A to choose another number, write it on both sides of a sheet of paper and hold it up.
Pupils now work out what number would be in their cell, write it on both sides of their sheet of paper and hold it up.
This can be repeated with different pupils deciding on the number for their cell.

◊ Questions can be posed in a class discussion, for example:

  - Suppose the number in Julie and Asif's cell is 20.
    What number is in your cell, Peter?
    What number will be in Jenny's cell?
  - What number do we need to put in cell A so that
    the number in cell F is 100?
  - Find a number for cell A so that the number in cell K is negative.
  - What happens if the 'across' and 'down' rules change places?

Ask pupils to explain how they worked out their answers. You could introduce the idea of an 'inverse' and encourage more confident pupils to use this word in their explanations.

Square grids

◊ Point out that all grids in the unit are square grids.

◊ One teacher presented unfinished grids on an OHP transparency and asked for volunteers to fill in any empty square. She found that less confident pupils chose easy squares to fill in while 'others with more confidence chose the hardest, leading to class discussion, and the idea of a "diagonal" rule came out naturally.'

◊ In one school, the class looked at rules in every possible direction as shown in the diagram.

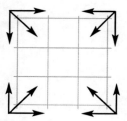

◊ In discussion, bring out the fact that there are different ways to calculate a number in a square depending on your route through the grid.

A1 In part (b), make sure pupils realise that the diagonal rule fits *any* position on these grids and not just those on the leading diagonal.
In part (c), some pupils may continue to use the '+ 6' and '+ 2' rules here. Discuss why using the diagonal rule '+ 8' could be used to give the same result.

A3 Some could consider what the diagonal rule is if the across and down rules are '+ a' and '+ b' or '– x' and '– y'.

A4 Some pupils can look for all possible pairs, introducing the idea of an infinite number of pairs of the forms

'+ a' and '+ (11 – a)' in part (a)

'+ b' and '+ (4 – b)' in part (b)

Ⓑ Grid puzzles (p 170)

◊ Pupils need not complete the grids to solve the puzzles; just find the missing number or rules.

◊ Encourage pupils to use the word 'inverse' when describing their methods.

B3 Pupils could solve puzzle (c) using trial and improvement or possibly by the following more direct method.

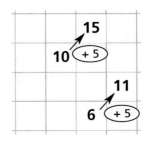

The rule in this diagonal direction (↗) is '+5' so the number in the top row directly above the 6 is 10 + 5 = 15.
Now the down rule is easily found to be '−3' and the across rule is '+2'.

As a possible extension, pupils could make up their own puzzles.

*B5 Pupils can devise their own methods to solve these. Ask them to explain them to you or each other. Some may use a direct method such as the one above. Others may devise systematic trial and improvement methods.

Ⓒ Algebra on grids (p 171)

◊ The first grid on the page uses addition only. After discussing it, you may wish pupils to try questions C1 to C3 where the rules are restricted to addition. Then discuss the second grid at the top of the page before moving on to question C4.

You may find a number line is helpful in getting across the fact that $^-4 + 7$ is equivalent to + 3, for example.

Emphasise that the expressions in the grid show how to find any number on the grid *directly* from the top left corner.

In the second grid, pupils who suggest '$h - 9$' for the square below '$h - 2$' are possibly thinking of '$h - 2 + 7$' as '$h - (2 + 7)$'. Discussion of numerical examples may help to clear up any confusion.

Ensure that pupils understand, for example, that '$n + 7 - 4$' gives the same result as '$n - 4 + 7$'.

Ⓓ Grid investigations (p 173)

D3 This provides an opportunity for pupils to choose to use algebra for themselves to explain their findings.

D4 After pupils have investigated opposite corners, draw their conclusions together in a discussion that leads to the algebraic ideas in section E.

E Using algebra (p 174)

T

This section follows on directly from the pupils' investigations in section D.

Extend your discussion to consider how to simplify expressions such as $n - 3 + n - 4$ and $n + n + 2 + n - 3$.

As an extension, pupils could find and prove that the opposite corners total on a 3 by 3 grid is 2 times the centre number or that the diagonals total is 3 times the centre number.

F Using multiplication (p 176)

Pupils simplify expressions such as $2n \times n$ and $3x \times 2x$.

F1 Pupils have the opportunity to use algebra to explain their answer, possibly stating that $3(n + 2) \neq 3n + 2$.

Encourage pupils to give explanations in a variety of ways using numbers, diagrams and/or algebra.

F2 Ask pupils if it will always be possible to make a grid with two '×' rules.

F5 Pupils could substitute various values for y to convince themselves that Tom is correct.

F12 Pupils can show algebraically that, on these grids, adding pairs of numbers in opposite corners does not produce the same results. Make sure they find that multiplying gives equal results and that they explain this algebraically.

A Square grids (p 169)

A1 (a) The pupil's grids

(b) The rule is '+ 8', with the pupil's explanations.

(c)

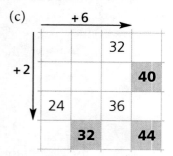

A2 (a) (i)

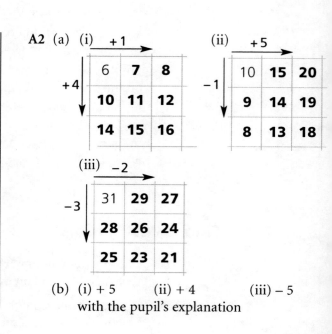

(b) (i) + 5　　(ii) + 4　　(iii) − 5
with the pupil's explanation

A3 (a) $+17$ (b) $+5$

A4 The pupil's pairs of rules, for example,

   (a) Across '$+1$', down '$+10$', or across '$+3$', down '$+8$'

   (b) Across '$+1$', down '$+3$', or across '-1', down '$+5$'

   Rules in (a) are of the form $+a, +(11-a)$

   Rules in (b) are of the form $+a, +(4-a)$

Ⓑ **Grid puzzles** (p 170)

B1 (a) 30   (b) 43   (c) 95

B2 (a) -3 (b) $+21$ (c) -5

B3 (a) Across '$+7$', down '-3'

   (b) Across '-2', down '$+11$'

   (c) Across '$+2$' down '-3'

B4 (c) is usually the most difficult. The pupil's reasons

*__B5__ (a) Across '-1', down '$+5$'

   (b) Across '-3', down '-4'

   (c) Across '$+4$', down '-1'

   (d) Across '$+6$', down '-2'

Ⓒ **Algebra on grids** (p 171)

C1 (a)

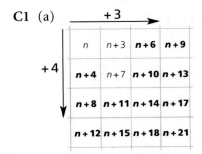

   (b)

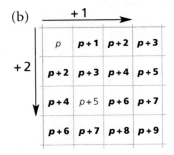

(c)

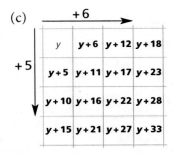

C2 (a) 121   (b) 109   (c) 133

C3 (a) Across '$+6$', down '$+5$'

   (b) Across '$+5$', down '$+1$'

   (c) Across '$+3$', down '$+10$'

C4 (a)

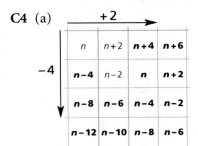

   (b)

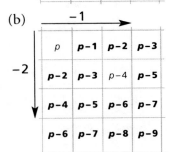

   (c)

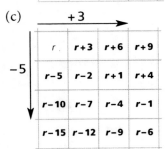

C5 (a) 56   (b) 59   (c) 56

C6 (a) $f+11$ (b) $y+12$ (c) $x+6$

   (d) $z-10$ (e) $p+1$ (f) $m-3$

   (g) $q+3$ (h) $w-14$ (i) $h+5$

C7

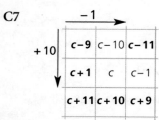

-1

$c-9$	$c-10$	$c-11$
$c+1$	c	$c-1$
$c+11$	$c+10$	$c+9$

$+10$

C8 (a)

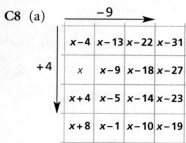

-9

$x-4$	$x-13$	$x-22$	$x-31$
x	$x-9$	$x-18$	$x-27$
$x+4$	$x-5$	$x-14$	$x-23$
$x+8$	$x-1$	$x-10$	$x-19$

$+4$

(b)

$+5$

$r+4$	$r+9$	$r+14$	$r+19$
$r+2$	$r+7$	$r+12$	$r+17$
r	$r+5$	$r+10$	$r+15$
$r-2$	$r+3$	$r+8$	$r+13$

-2

(c)

-2

$y+4$	$y+2$	y	$y-2$
$y+1$	$y-1$	$y-3$	$y-5$
$y-2$	$y-4$	$y-6$	$y-8$
$y-5$	$y-7$	$y-9$	$y-11$

-3

C9 (a) Across '$+4$', down '-1'

(b) Across '-5', down '-3'

(c) Across '-2', down '-1'

C10 (a) $+5-x$ or equivalent

(b) $+10-n$ or equivalent

Ⅾ **Grid investigations** (p 173)

D1 (a) $37+13=50$

(b) The pupil's grids

(c) For each grid, the opposite corners totals are the same – this result could be explained using algebra.

D2 The pupil's investigation

D3 (a)

Opposite corners table	
Top left number	Opposite corners total
2	10
3	12
4	14
10	26

(b) The pupil's grids and results

(c) 'The opposite corners total is (the top left number × 2) + 6' or '… (the top left number + 3) × 2' or '… $2n+6$' or equivalent.

(d) 206

D4 The pupil's investigation

Ⅿ **Using algebra** (p 174)

E1 Grid P

(a)

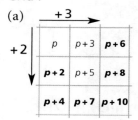

$+3$

p	$p+3$	$p+6$
$p+2$	$p+5$	$p+8$
$p+4$	$p+7$	$p+10$

$+2$

(b) Both pairs of corners add up to $2p+10$.

(c) Yes (d) 210

Grid N

(a)

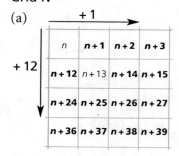

$+1$

n	$n+1$	$n+2$	$n+3$
$n+12$	$n+13$	$n+14$	$n+15$
$n+24$	$n+25$	$n+26$	$n+27$
$n+36$	$n+37$	$n+38$	$n+39$

$+12$

(b) Both pairs of corners add up to $2n+39$.

(c) Yes (d) 239

Grid T

(a)

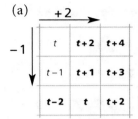

(b) Both pairs of corners add up to $2t + 2$.

(c) Yes   (d) 202

E2 A and I, B and G, C and E, D and H

E3 The pupil's explanation, for example, the expressions have the same value for **one** value of n but this does not mean that the expressions have the same value for **all** values for n.

E4 (a) $2p + 6$ (b) $2y + 9$ (c) $3q + 8$
    (d) $3t + 4$ (e) $2x + 1$ (f) $3r + 6$
    (g) $2w - 9$ (h) $2j - 1$ (i) $3h - 2$

E5 The pupil's investigation

E6 Pairs that add to give $+ 15$, for example '$+ 8$' and '$+ 7$', '$+ 16$' and '$- 1$'

$\mathbb{F}$ Using multiplication (p 176)

F1 (a) There is no single right answer. They have taken different routes through the grid.

     (b) No, it is not possible – the pupil's explanation

F2 (a)

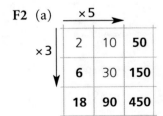

     (b) The pupil's explanation

F3 (a) No   (b) Yes   (c) No
    (d) Yes   (e) Yes

F4 (a)

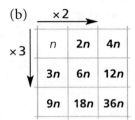

     (b)

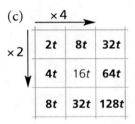

     (c)

$\times 2$ $\times 4 \rightarrow$

$2t$	$8t$	$32t$
$4t$	$16t$	$64t$
$8t$	$32t$	$128t$

F5 Tom is correct. The pupil's reasons

F6 (a) $2a$ (b) $2a^2$ (c) $6a^2$ (d) $9a^2$
    (e) $5x^2$ (f) $10x^2$ (g) $25x^2$ (h) $36x^2$

F7 A and D, B and J, C and F, E and H, G and I

F8 (a) $6p^2$ (b) $5p$ (c) $6p$ (d) $3t^2$
    (e) $15t^2$ (f) $15t$ (g) $5t^2$ (h) $24x^2$
    (i) $24x^2$ (j) $20y^2$ (k) $15y^2$ (l) $24y^2$

F9 The pupil's explanation

F10 (a) $12p$ (b) $7p$ (c) $12p^2$
     (d) Cannot be simplified
     (e) $p + 7$ (f) $2p + 7$ (g) $2p + 1$
     (h) Cannot be simplified
     (i) $8m^2$ (j) $9m$
     (k) Cannot be simplified   (l) $7m + 2$

F11 (a) $2x - 6$ (b) $16 + 3y$ (c) $6w^2$
     (d) $4u + 4$ (e) Cannot be simplified
     (f) $18p^2$ (g) Cannot be simplified
     (h) $2r - 4$ (i) $4s - 3$

F12 The pupil's investigation

What progress have you made? (p 178)

1 (a) Across '– 5', down '– 3'
   (b) Across '+ 2', down '– 3'

2 (a) $n + 7$ (b) $p + 6$ (c) $y - 7$
   (d) $t - 3$ (e) $2h + 3$ (f) $3w - 11$

3 (a)

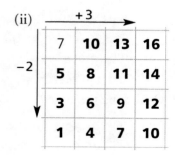

   (b) The pupil's explanation, for
      example, the expressions in one
      diagonal are all the same (n) so the
      numbers will be the same.
   (c) $2n$ (d) 12
   (e) The pupil's investigation

4 (a) t^2 (b) $5g^2$ (c) $12n^2$ (d) $10y^2$

Practice booklet

Sections A and B (p 56)

1 (a) (i)

+5		
7	**12**	**17**
5	**10**	**15**
3	**8**	**13**

-2

   (ii)

+3			
7	**10**	**13**	**16**
5	**8**	**11**	**14**
3	**6**	**9**	**12**
1	**4**	**7**	**10**

-2

(iii)

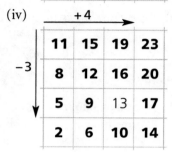

+2			
5	**7**	**9**	**11**
11	**13**	15	**17**
17	**19**	**21**	**23**
23	**25**	**27**	**29**

+6

(iv)

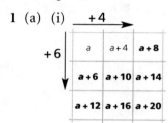

+4			
11	**15**	**19**	**23**
8	**12**	**16**	**20**
5	**9**	13	**17**
2	**6**	**10**	**14**

-3

   (b) (i) + 3   (ii) + 1   (iii) + 8   (iv) + 1

2 (a) + 13   (b) + 2

Section C (p 56)

1 (a) (i)

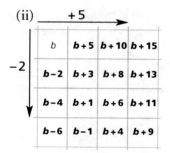

+4		
a	$a+4$	$a+8$
$a+6$	$a+10$	$a+14$
$a+12$	$a+16$	$a+20$

+6

   (ii)

+5			
b	$b+5$	$b+10$	$b+15$
$b-2$	$b+3$	$b+8$	$b+13$
$b-4$	$b+1$	$b+6$	$b+11$
$b-6$	$b-1$	$b+4$	$b+9$

-2

   (b) (i)   120   (ii)   109
   (c) (i)   80   (ii)   91

2 (a) Across '+ 3', down '+ 4'
   (b) Across '+ 5', down '+ 3'
   (c) Across '– 1', down '+ 4'

3 (a) $t + 11$ (b) $a + 7$ (c) $q + 6$

   (d) $p + 4$ (e) $x + 10$ (f) $y + 2$

   (g) $s - 4$ (h) $v - 5$ (i) $b + 6$

   (j) $a - 4$ (k) $f - 16$ (l) $c + 2$

   (m) $d - 11$ (n) $g - 2$ (o) $h - 14$

Sections D, E and F (p 57)

1 (a)

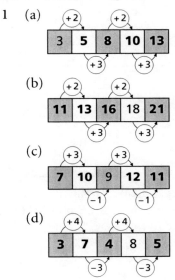

   (b)

   (c)

   (d)

2 (a) $+4$ (b) $+7$ (c) $+8$ (d) $+4$

3 The pupil's pairs of rules, for example,
   Top '$+ 1$', bottom '$+ 3$'
   Top '$+ 3$', bottom '$+ 1$'
   Top '$+ 2$', bottom '$+ 2$'
   Top '$+ 5$', bottom '$- 1$'

4 (a)

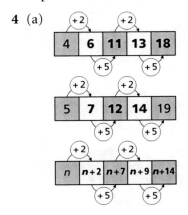

(b)

End total table	
First number	End total
3	20
4	22
5	24
n	$2n + 14$

(c) 'The end total is
   2 times the first number add 14' or
   '… (the first number × 2) + 14' or
   '… (the first number + 7) × 2' or
   '… $2n + 14$ where n is the first
   number' or equivalent.

(d) 30

5 The pupil's investigation

6 A and D, C and I, E and G

7 (a) $2p + 3$ (b) $3y + 9$ (c) $2q + 10$

   (d) $2t + 3$ (e) $2x + 2$ (f) $3r + 6$

   (g) $2w - 13$ (h) $2j - 3$ (i) $2h - 6$

8 (a)

Grid total table	
Top left	Grid total
2	63
3	72
4	81

(b) 'The grid total is 9 times the top left
   number add 45' or
   '… (the top left number × 9) + 45'
   or
   '… (the top left number + 5) × 9' or
   '… $9n + 45$ where n is the top left
   number' or equivalent.

(c) 495

9 The pupil's investigation

10 (a) $6n^2$ (b) $35p$ (c) $4m - 3$

   (d) Cannot be simplified

   (e) $3a^2$ (f) $s + 5$

24 Constructions

Essential
Set square, compasses, sheet 139, squared paper
Practice booklet page 60

A Right angles (p 179)

Set square

B From a point to a line (p 181)

Set square, sheet 139, squared paper

T

The fact that the shortest distance is perpendicular to the line appears obvious when the point is above a horizontal line. You could start with this: and rotate to this:

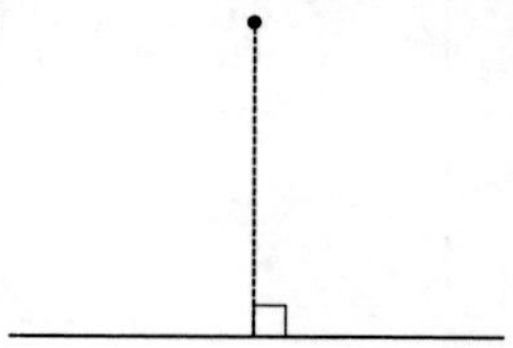

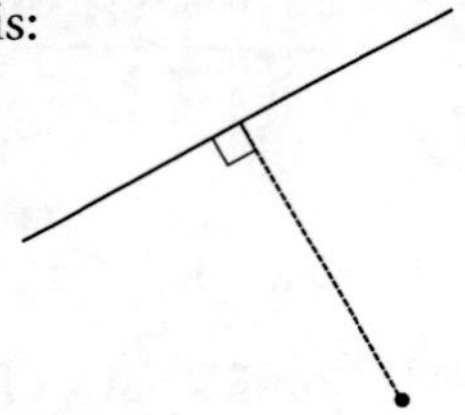

C Constructions with ruler and compasses (p 182)

Compasses

Ask pupils why the constructions work. The most accessible explanations are based on reflection symmetry.

A Right angles (p 179)

A1 *a* and *h*, *b* and *d*, *e* and *g*

A2 (a) North   (b) North-east
     (c) North-east

A3 The pupil's drawing

A4 *a* and *b*, *c* and *j*, *d* and *e*, *i* and *l*.

B From a point to a line (p 181)

B1 The pupil's mirror images

B2 The pupil's diagram; (4, 2)

B3 The total of the three distances is the same, whatever the position of P.

ℂ Constructions with ruler and compasses (p 182)

C1 The three lines (possibly extended) all go through a point.

C2 They all go through a point.

C3 They meet at the midpoint of AC.

C4 They all go through a point.

C5 The angle bisectors are at right angles.

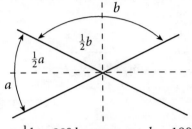

$\frac{1}{2}a + \frac{1}{2}b = 90°$ because $a + b = 180°$.

C6 The angles of the large triangle are equal to those of the original triangle.

What progress have you made? (p 185)

1 The pupil's construction

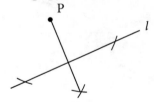

2 The pupil's construction

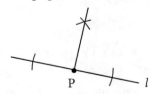

3 The pupil's construction

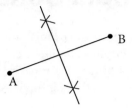

4 The pupil's construction

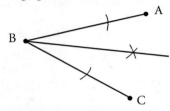

Practice booklet

Section C (p 60)

1 The pupil's construction of an angle of 45°

2 (a) 60°

(b) The pupil's construction of an angle of 30°

3 The three points A, B, C are on a straight line.

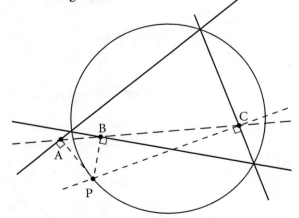

25 Comparisons 1

This unit introduces median and range and uses these to compare sets of data. Pupils generate their own data in a reaction time experiment and use it to make comparisons. Other data handling projects are suggested and there is advice on writing up.

T   p 186 **A** Comparing heights

p 188 **B** Median

p 190 **C** Range

T   p 192 **D** How fast do you react?   Making and using a simple reaction timer to generate data for comparison

p 193 **E** Summarising data   Using extremes, median and range as a summary

T   p 194 **F** Writing a report   Discussing a specimen report

T   p 196 **G** Projects   Activities for the class, and for pairs or groups, generating data for comparison

p 196 **H** Puzzles and problems   Problems based on median and range

Essential	Optional
Sheet 141	Sheets 142–145
Practice booklet pages 61 to 63	

A Comparing heights (p 186)

◊ The discussion here is intended to be open, with no particular method of comparison preferred. The important thing is to give reasons for decisions.

The two questions (about the picture and about the dot plots) can be given to pairs or small groups to consider before a general discussion.

There may not always be a clear-cut answer.

◊ Pupils may suggest finding the mean. You could offer data where the mean of one group is greater than the mean of another, but only because of one extreme value (for example, 190, 150, 150, 150 and 159, 159, 159, 159). Work on the mean and discussion of which 'average' is more appropriate are dealt with later in the course.

B Median (p 188)

◊ You can introduce the median using the pupils' own heights. They will need to know their heights or will need to measure them.

Start by getting an odd number of pupils to stand in order of height. Emphasise that it is not the middle person who is the median of the group, but that person's height. It is a good idea to use 'median' only as an adjective at first, for example 'median height', 'median age'.

Then do the same with an even number of pupils.

To emphasise the value of the median as a way of comparing data, you may wish to carry out this activity with two separate groups (boys and girls or sides of the class).

C Range (p 190)

◊ The range can be shown practically with a group of pupils. Ask the group to stand in height order. Ask the tallest and shortest to stand side by side. Measure the difference between their heights.

D How fast do you react? (p 192)

Pupils work in pairs.

Sheet 141

◊ In addition to comparing performance within each pair, pupils could compare left and right hands. For homework they could compare themselves with an adult.

◊ In one class pupils felt that they were getting clues from twitching fingers just before the ruler was dropped, so a card was used to cover the fingers.

◊ You may wish to work through section F (on report writing) first and get pupils to write up their work on reaction times. If so, you will need to explain the diagrams in section E as well.

E Summarising data (p 193)

The diagram in the pupil's book is a simplified version of the 'box and whisker' diagram, which shows the quartiles as well.

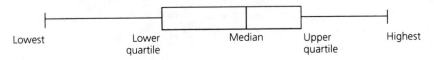

Lowest   Lower quartile   Median   Upper quartile   Highest

Either type of diagram may be arranged vertically.

Make sure pupils align the diagrams for each question correctly on a common scale, so that they can be used to make comparisons.

F **Writing a report** (p 194)

T

The specimen report is intended to help pupils write their own reports. It can be used for either group or class discussion

◊ You may need to make it clear that the fastest person is the one with the shortest time, and vice versa.

G **Projects** (p 196)

Each project generates data for making comparisons in a short written report.

The Argon Factor

> Optional: Sheets 142 and 143 (preferably on OHP transparencies), sheet 144

T

◊ There are two tests: mental agility and memory.

◊ In the mental agility test, give the pupils one minute to memorise the shapes and numbers. (They are best shown on an OHP.)

Tell pupils 'You will be given 5 seconds to answer each question. Questions will be read twice.'

1 What number is inside the circle?
2 What number is inside the pentagon?
3 What number is inside the first shape?
4 What shape has the number 17 in it?
5 What shape has the number 29 in it?
6 What number is inside the middle shape?
7 What number is inside the last shape?
8 What shape is in the middle?
9 What is the shape before the end one?
10 What number is in the second shape?
11 What shape is the one after the one with 21?
12 What number is inside the fourth shape?
13 What number is in the shape just right of the triangle?
14 What shape is two to the left of the kite?
15 What number is in the shape two after the circle?

'Argon factor is excellent as a class activity and well worth the time spent on it.'

◊ In the memory test, give the pupils two minutes to remember the pictures and details of the four people on sheet 143.

Then give them ten minutes to write answers to the 20 questions on sheet 144.

◊ Discuss with the class what comparisons can be made from the scores. Suggestions include these.

  • Do people remember more about their own sex?
  • Which test did the class do better on? (Remember that the number of questions is not the same.)

Other projects

Optional: For 'Tile pattern' (see below), sheet 145

Another project idea is given on p 60 ('Handwriting size').

Another possibility is 'Tile pattern', for which sheet 145 is needed. This is suitable for an individual pupil or a pair. The pupil(s) carrying out the project cut out the 16 tiles and put them in an envelope. They decide which groups of people are to be compared (for example, children and adults). Each 'subject' is then timed making the pattern shown on the sheet.

⊞ Puzzles and problems (p 196)

Some questions here and in the practice leaflet involve finding the median of a frequency distribution. This topic is developed further at a later stage.

Ⓑ Median (p 188)

B1 (a) 159 cm (b) 156 cm (c) 154.5 cm

B2 (a) 11 (b) 154 cm

B3 A Girls 136 cm; boys 153 cm
  The boys are taller.

  B Girls 149.5 cm; boys 143 cm
  The girls are generally taller.

  C Girls 143 cm; boys 149 cm
  The boys are generally taller but the girls' heights are well spread out.

  D Girls 149 cm; boys 140 cm
  The girls are generally taller but there are some short girls and one tall boy.

  E Girls 139 cm; boys 149 cm
  The boys are generally taller but there is a tall girl and some short boys.

  F Girls 152 cm; boys 145 cm
  The girls are generally taller.

B4 (a) 152 cm (b) 151 cm
  (c) 153 cm (d) 152.5 cm

B5 (a) 69 kg (b) No change
  (c) No change (d) Up by 1 kg

B6 (a) 52, 54, 58, 60, 63, 65, 70
  (b) 60 kg

B7 (a) 152 cm (lengths in order are 139, 148, 152, 156, 161 cm)

(b) 36 kg (weights in order are 26, 29, 31, 34, 38, 39, 40, 45 kg)

B8 Boys have the greater median weight (2.7 kg); the girls' median is 2.65 kg.

ℂ Range (p 190)

C1 (a) A 13 cm   B 11 cm   C 18 cm

(b) C

(c) B

C2 (a) 13 minutes   (b) 4 minutes

(c) 9 minutes

C3 (a) Herd B

(b) Herd A, because it has the greatest range

(c) Herd C, because it has the smallest range

C4 (a) Median 28, range 12

(b) Median 85, range 29

(c) **Nicky** and **Carol** both had high scores, but **Nicky**'s scores were the more spread out of the two.

(d) Nicky's scores and **Martin**'s scores were both spread out, but **Nicky** had the higher scores of the two.

(e) **Paul** and **Martin** were both bad players because they had **low** median scores.

(f) Paul was a consistent player because the range of his scores was **low**.

C5 (a) Northern: median 12 m, range 7 m
Southern: median 14.5 m, range 11 m

(b) Southern trees are taller. Their heights are more spread out.

C6 (a) Machine A: median 500 g, range 26 g
Machine B: median 498 g, range 5 g
Machine C: median 502 g, range 6 g
Machine D: median 514 g, range 25 g

(b) Machine B

(c) Machine D

(d) Machine C

(e) (Machine A) Inconsistent: it both underfilled and overfilled packs

(f) Machine C. It usually put enough in a pack to avoid complaint without being too generous to the customer.

𝔼 Summarising data (p 193)

E1 The pupil's diagrams

E2 (a) (i)

(ii) (iii)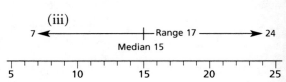

(b) Linford is consistently quick. He has the lowest median time and the smallest range.
Jules has some quick times, but is the least consistent, as shown by the large range.

ℍ Puzzles and problems (p 196)

H1 The missing age could be 5 or 48. If it is 5, the median is 16; if it is 48, the median is 20.

H2 (a) 20

(b) (i) 11, 15 or 18   (ii) 20
(iii) 24, 25 or 27

H3 (a) 2 people   (b) 3 people

*****H4** 43, 46, 55, 68, 72 and 48, 58, 62, 79

What progress have you made? (p 197)

1 Right hand: median 14, range 6
Left hand: median 18, range 16
Pat is faster and less variable using his
right hand.

Practice booklet

Sections A, B and C (p 61)

1 (a) 138 g   (b) 140 g   (c) 207 g

2 (a)

   (b) 132 g

3 (a) Barcelona 21°C
Birmingham 22°C

   (b) Birmingham

   (c) Barcelona 12 degrees
Birmingham 8 degrees

Sections C and E (p 62)

1 The lengths are
A 27 mm   B 33 mm   C 36 mm
D 40 mm   E 56 mm   F 53 mm
Median length 38 mm
Range of lengths 29 mm

2 (a) Median 85 kg, range 46 kg

   (b) Median 7.8 m, range 3.3 m

   (c) Median 1°C, range 10 degrees

   (d) Median 37.7°C, range 2.8 degrees

3 The pupil's comparison, such as
'The grey squirrels weigh more, because
the median weight of the red squirrels is
293 g, and the median of the greys is
599 g. The greys are also more varied in
weight, as the range of the reds is 25 g,
but the range of the greys is 112 g.'

4 On the whole, Jo is faster (the medians
are 80.1 s and 81.7 s). Jo is also more
consistent (the ranges are 3.4 s and 8.3 s).
But Jay has had the fastest single
run (76.2 s).

5 Jo

   78.3 ◄ Range 3.4 ├──────► 81.7
             Median 80.1

   Jay

   76.2 ◄── Range 8.3 ──┼──────► 84.5
             Median 81.7

Section H (p 63)

1 24

2 (a) 23   (b) 25

3 $23\frac{1}{2}$

4 2

5 17

26 Further areas

T p 198 **A** Area of a parallelogram

T p 201 **B** Area of a triangle

T p 203 **C** Area of a trapezium

> **Essential**
> Angle measurer, set square
> **Practice booklet** pages 64 to 67

A Area of a parallelogram (p 198)

The aim is for pupils to justify for themselves the formula for the area of a parallelogram and to gain confidence in using the correct dimensions. The work provides an opportunity for pupils to practise accurate drawing.

> Angle measurer, set square

◊ Drawing the parallelogram from the sketch gives an opportunity to revise work on parallel lines and constructions. The size of the obtuse angle needs to be found (110°).

◊ These points should emerge from dissecting the parallelogram.
- There are essentially two ways of dissecting the parallelogram, cutting at right angles to the longer side or at right angles to the shorter side.
- These two ways justify the two ways of measuring base and height for use in the formula.

The parallelogram has an area of about 66 cm². It can be made into a rectangle 7 cm by 9.4 cm or 10 cm by 6.6 cm.

A set square should be used as a guide when measuring perpendicular to a side.

◊ The approach above breaks down in the case of an 'overhanging' parallelogram, which will make a rectangle if cut at right angles to its longer side but not its shorter side. However, the approach in question A3 shows that the 'base × height' formula still applies when the shorter side is the base.

⅛ **Area of a triangle** (p 201)

◊ The main difficulty is identifying the correct lengths to multiply by. You could ask 'What parallelogram could the triangle be half of?'
There are of course three for every triangle:

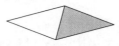

The next question is
'Which of the parallelograms do we have the measurements for?'

Again, a set square should be used as a guide when measuring perpendicular to a 'base'.

ℂ **Area of a trapezium** (p 203)

◊ Draw a trapezium on the board, with the lengths of its parallel sides and the distance between them given. Pupils could work in pairs or small groups to try to find its area.

Pupils may all approach the problem in the same way, but you could get variations. Pupils who solve the problem quickly can be asked to try to find a different way of dissecting the trapezium.

◊ After discussing the approaches used, you could refer to some of the variations below.

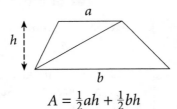

$$A = \tfrac{1}{2}ah + \tfrac{1}{2}bh$$

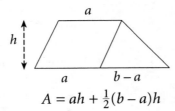

$$A = ah + \tfrac{1}{2}(b-a)h$$

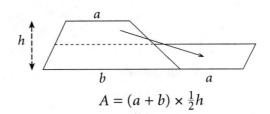

$$2A = (a+b)h$$

$$A = (a+b) \times \tfrac{1}{2}h$$

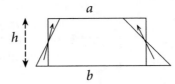

$$A = h \times \frac{(a+b)}{2}$$

Some people find it easier to think of the formula for a trapezium as
'find the mean of the lengths of the parallel sides by adding them and dividing by 2,
then multiply by the distance between them'
For this interpretation the formula can be written as

$$A = \frac{(a + b)}{2} h$$

The more complicated questions (C5 and C6) could be used to practise efficient use of a calculator's memory.

All the answers based on measurements can be expected to differ slightly from these.

Ⓐ Area of a parallelogram (p 198)

A1 About 69 cm²

A2 About 48 cm²

A3 P and R are equal in area,
because A + P + B = R + A + B

A4 (a) 10 cm²   (b) 22.62 cm²
     (c) 13.02 cm²   (d) 29.11 cm²

A5

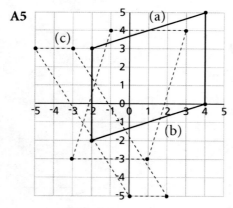

(a) 30 square units

(b) 28 square units

(c) 16 square units

A6 45.72 cm²

A7 $a = 2.4\,\text{m}$ $b = 1.8\,\text{m}$ $c = 2.5\,\text{m}$
    $d = 2.1\,\text{m}$ $e = 3.0\,\text{m}$

A8 $a = 3.0\,\text{m}$ $b = 5.0\,\text{m}$ $c = 7.0\,\text{m}$

*A9 The area can be found by subtracting pieces from a square.

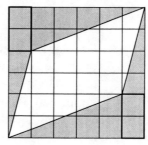

Area 18 square units

Ⓑ Area of a triangle (p 201)

B1 (a) 13.65 cm²   (b) 11.6 cm²
    (c) 10.2 cm²   (d) 3.22 cm²

B2 The pupil's answers may differ slightly because of measurement variations.

(a) Base = 5.0 cm and height = 6.0 cm or
Base = 6.1 cm and height = 4.9 cm or
Base = 7.1 cm and height = 4.2 cm
Area = 15.0 cm²

(b) Base = 5.3 cm and height = 7.5 cm or
Base = 7.6 cm and height = 5.2 cm or
Base = 8.8 cm and height = 4.5 cm
Area = 19.8 cm²

(c) Base = 5.0 cm and height = 7.2 cm or
Base = 7.6 cm and height = 4.7 cm or
Base = 10.3 cm and height = 3.5 cm
Area = 17.9 cm²

B3

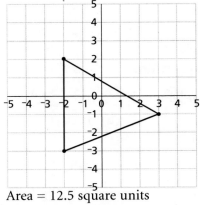

Area = 12.5 square units

B4 (a) $61.6\,cm^2$ (b) $43.5\,cm^2$
    (c) $29.0\,cm^2$ (d) $52.2\,cm^2$

B5 (a) $5.5\,cm$ (b) $5.2\,cm$ (c) $6.0\,cm$

ℂ **Area of a trapezium** (p 203)

C1 (a) $17.0\,cm^2$ (b) $14.0\,cm^2$ (c) $31.2\,cm^2$

C2 $17.1\,cm^2$

C3 The pupil's answers may differ slightly because of measurement variations.
    (a) $12.3\,cm^2$ (b) $23.9\,cm^2$

C4 $22.6\,m^2$

C5 (a) $178.3\,m^2$ (to 1 d.p.)
    (b) £3298 to the nearest £

C6 $4411.5\,m^2$

C7 $11.76\,m^2$

C8 $218.2\,m^2$

*C9 (a) $6.4\,m$ (b) $10.1\,m$

What progress have you made? (p 206)

The pupil's answers to 1 and 4 may differ slightly because of measurement variations.

1 $7\,cm^2$

2 $11.7\,m^2$

3 $6.0\,m$

4 $4.6\,cm^2$

5 $2.1\,cm^2$

6 $42.75\,m^2$

Practice booklet

Section A (p 64)

1 (a) $15\,cm^2$ (b) $7.0\,cm^2$
  (c) $22.4\,cm^2$ (d) $4.7\,cm^2$
  (e) $15.75\,cm^2$ (f) $17.5\,cm^2$
  (g) $25.5\,cm^2$ (h) $12.6\,cm^2$

2 (a) $12\,cm$ (b) $4.5\,cm$
  (c) $4\,cm$ (d) $6\,cm$

Section B (p 65)

1 $12\,cm^2$: A B D E G L M
  $18\,cm^2$: C H I J
  $24\,cm^2$: F K N

2 (a) $11.76\,cm^2$ (b) $11.96\,cm^2$
  (c) $10.75\,cm^2$ (d) $11.22\,cm^2$
  (e) $14.28\,cm^2$ (f) $19.74\,cm^2$

3 The pupil's answers may differ slightly because of measurement variations.

  (a) Triangle $1.4\,cm^2$, rectangle $2.8\,cm^2$
      Total $4.2\,cm^2$

  (b) Both parallelograms $2.8\,cm^2$
      Total $5.6\,cm^2$

  (c) Triangles (left to right) $3.0\,cm^2$,
      $4.0\,cm^2$, $2.6\,cm^2$
      Total $9.6\,cm^2$

  (d) Triangle $17.2\,cm^2$
      Parallelogram $2.3\,cm^2$
      Shaded area $14.9\,cm^2$

  (e) Triangle $15.8\,cm^2$, rectangle $4.3\,cm^2$
      Shaded area $11.5\,cm^2$

Section C (p 67)

1 $72\,cm^2$: A F G
  $48\,cm^2$: B C
  $60\,cm^2$: D E

2 $198.4\,m^2$

Review 3 (p 207)

1 $a = 18.5\,\text{m}$ $b = 15.5\,\text{m}$

2 (a) $^-11$ (b) 5   (c) $^-5$ (d) $^-5$
   (e) 11 (f) 11 (g) 5   (h) $^-5$

3 $\frac{1}{25}$

4 (a) $\begin{bmatrix} 6 \\ -2 \end{bmatrix}$

   (b) (12, 1) (16, 2) (15, $^-2$)

   (c) $\begin{bmatrix} -12 \\ 4 \end{bmatrix}$

5 (a) 9   (b) $^-5$ (c) 8   (d) $^-4$

6 (a) $2x + 11$ (b) $2y + 1$ (c) $3a - 9$
   (d) $4b^2$ (e) $30c$ (f) $18d^2$

7 The perpendicular bisectors all go through a point (the centre of the circle).

8 (a) 171 mm   (b) 21 mm
   (c) Overall the boys' handspans are larger than the girls'.
      The boys' handspans are less spread out than the girls'.

9 (a) 35.36 cm²   (b) 20.68 cm²

10 (a) RR, RB, RG, RY
       BR, BB, BG, BY
       GR, GB, GG, GY
       YR, YB, YG, YY
    (b) $\frac{9}{16}$

11 45.9 cm²

Mixed questions 3 (Practice booklet p 68)

1 (a) 15.12 cm²   (b) 12 cm²

2 (a) $^-3$ (b) 4   (c) 7
   (d) $^-15$ (e) $^-6$

3 (a) $n \rightarrow 2n^2$ (b) 98

4 (a) $\frac{3}{4}$ (b) $\frac{1}{2}$ (c) $\frac{1}{2}$
   (d) $\frac{1}{4}$ (e) $\frac{1}{8}$ (f) 0

5 A goes to (1, 4),
   B goes to ($^-4$, 5),
   C goes to ($^-1$, 0).

6 (a)

n	$n + 3$	$n + 6$
$n - 2$	$n + 1$	$n + 4$
$n - 4$	$n - 1$	$n + 2$

   (b) $9n + 9$ or $9(n + 1)$

7 (a)–(e) The pupil's constructions
   (f) (Regular) octagon

8 Median = 26 seats, range = 36 seats

9 1827 m²

10 $\frac{2}{5}$

11 No. The grey triangles have a smaller angle at the centre, so there is a smaller probability that the arrow will land in the grey.

12 (a) 0.6204   (b) 0.062 04 (c) 62.04

Inputs and outputs

This unit introduces pupils to equivalent expressions, such as
$3(a + 6) = 3a + 18$ and $\frac{4a - 20}{4} = a - 5$.

p 209 **A** Input and output machines	Using number machines	
p 210 **B** Shorthand rules	Using algebra to write rules	
p 212 **C** Evaluating expressions	Substituting into expressions such as $2(n + 5)$, $2n + 5$	
p 213 **D** Same but different!	Discussing equivalent rules such as $n \rightarrow 5(n - 3)$ and $n \rightarrow 5n - 15$	
p 214 **E** Equivalent expressions	Multiplying out brackets such as $6(n + 2) = 6n + 12$	
p 216 **F** Inventing puzzles	Using algebra to explain 'think of a number' puzzles	

Essential

Sheets 148 and 149

Practice booklet pages 71 to 73

Ⓐ **Input and output machines** (p 209)

◊ You could introduce the unit to the whole class, taking questions A1 and A2 orally.

A4 These questions should get pupils thinking about looking for shorter chains. By observing the number patterns (it may be helpful to add some consecutive values to the table) they should be able to find shorter chains.

Ⓑ **Shorthand rules** (p 210)

Pupils have met the fact that we can write $a \times 3$ as $3a$. Here the notation is slightly extended, in that we write $(c + 5) \times 4$ as $4(c + 5)$.

T

◊ In discussing the notation, bring out the fact that any letter can be used to represent an input number, and that when multiplication is involved, the number comes first and the multiplication sign is omitted.

Mention that because 4 × 3 = 3 × 4 we can (and do!) always put the number before the letter. You may want to use several examples like those in the introduction to ensure that pupils are secure with the notation. Include the fact that we can write *s* × 1 as simply *s*.

B2 What pupils write here can be used diagnostically. For example in part (a), if you see $c \rightarrow c + 12$, then it suggests the pupil is only multiplying the 4 by 3, rather than the whole of $c + 4$.

ℂ **Evaluating expressions** (p 212)

Sheet 148

'Pupils enjoyed the "Cover up" game, which provided strong reinforcement.'

◊ Your discussion should include how to use arrow diagrams to evaluate expressions with and without brackets. You could swap the operations in the example on the page to consider $5(p - 3)$.

◊ To break the ground for the 'Cover up' game on page 213, pupils could discuss different possible rules for a single input–output pair. For example, ask pupils in pairs to find as many rules as they can that fit $2 \rightarrow 5$.

𝔻 **Same but different!** (p 213)

'Spent lots of time on D1. It is very useful to refer to this exercise in later algebra work.'

This teacher-led section is for pupils to begin to consider when two rules are equivalent by looking at numerical examples.
You might deal with all the examples orally.

◊ Emphasise that the outputs need to be the same for any input.

◊ At the end of this section ask pupils what they have discovered. Hopefully they will have seen how the expressions in each pair of results relate to each other.

𝔼 **Equivalent expressions** (p 214)

This section follows on from the teacher-led section D.
The initial discussion should include examples that involve subtraction.

Sheet 149, one between two

◊ You could also use simple areas to show equivalence. For example

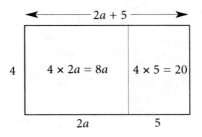

Include some examples where the number comes first in the brackets, for example $3(2 + n) = 6 + 3n$ or $3n + 6$.

Expression snap

The game provides a way of reinforcing pupils' understanding without too many tedious examples.

◊ You may wish to copy the sets of cards on to different coloured card. If groups of pupils have different coloured sets, it will help them sort the cards after each game.

F Inventing puzzles (p 216)

Pupils use the algebra they have learned to explain how 'think of a number' type puzzles work. Some of the puzzles give a fixed number, others end up with the starting number.

◊ Pupils have found it fun if the teacher reads out one puzzle to the class, and then asks individuals what their answers are. Amazingly (?) they are all the same. Pupils might then discuss in groups why this is so. One representative from each group may then give the group's explanation to the whole class.

Here is a way to write out an algebraic explanation.

Think of a number.	n
Multiply it by 2.	$2n$
Add on 6.	$2n + 6$
Divide by 2.	$\frac{2n + 6}{2} = n + 3$
Take off the number you first thought of.	$n + 3 - n = 3$
What is your answer?	3

Ⓐ Input and output machines (p 209)

A1 (a) $17 \xrightarrow{\times 2} 34 \xrightarrow{-5} 29$

(b) $6 \xrightarrow{\times 3} 18 \xrightarrow{-4} 14$

A2 (a) $2 \to \mathbf{2}$ (b) $6 \to \mathbf{7}$
$1 \to \mathbf{^{-}1}$ $9 \to \mathbf{8\frac{1}{2}}$
$\mathbf{8} \to 20$ $\mathbf{0} \to 4$
$\mathbf{0} \to {^{-}4}$ $3 \to 5\frac{1}{2}$

(c) $9 \to \mathbf{28}$ (d) $10 \to \mathbf{11}$
$4 \to \mathbf{18}$ $7 \to \mathbf{9.5}$
$5 \to 20$ $\mathbf{5} \to 8.5$
$\frac{1}{2} \to 11$ $\mathbf{72} \to 42$

A3 (a) $2 \to \mathbf{10}$ (b) $1 \to \mathbf{2}$
$8 \to \mathbf{28}$ $5 \to \mathbf{6}$
$\mathbf{10} \to 34$ $\mathbf{39} \to 40$
$\mathbf{5} \to 19$ $\mathbf{9} \to 10$

(c) $1 \to \mathbf{10}$ (d) $10 \to \mathbf{7}$
$2.1 \to \mathbf{15.5}$ $5 \to \mathbf{4.5}$
$\mathbf{7} \to 40$ $\mathbf{4} \to 4$
$\mathbf{1.1} \to 10.5$ $\mathbf{8.4} \to 6.2$

***A4** (a) $\times 3, +4$

(b) $+1$

(c) $\times 5, +5$ or $+1, \times 5$

(d) $\div 2, +2$ or $+4, \div 2$

Ⓑ Shorthand rules (p 210)

B1 (a) $a \xrightarrow{\times 3} 3a \xrightarrow{-2} 3a-2$

(b) $a \to 3a-2$

B2 (a) $c \to 3(c+4)$ (b) $p \to 2p+4$

B3 A is correct.

B4 (a) $a \to 5a-3$ (b) $a \to 5(a-3)$
(c) $w \to \dfrac{w+7}{2}$ (d) $w \to \dfrac{w}{2}+7$

B5 (a) $s \xrightarrow{\times 4} 4s \xrightarrow{+5} 4s+5$

(b) $t \xrightarrow{-5} t-5 \xrightarrow{\div 3} \dfrac{t-5}{3}$

(c) $w \xrightarrow{\times 5} 5w \xrightarrow{-7} 5w-7$

(d) $x \xrightarrow{\div 4} \dfrac{x}{4} \xrightarrow{-1} \dfrac{x}{4}-1$

(e) $y \xrightarrow{\times 7} 7y \xrightarrow{+3} 7y+3$

(f) $z \xrightarrow{+5} z+5 \xrightarrow{\times 2} 2(z+5)$

B6 $f \xrightarrow{\times 2} 2f \xrightarrow{+3} 2f+3 \xrightarrow{\div 5} \dfrac{2f+3}{5}$

The rule is $f \to \dfrac{2f+3}{5}$.

B7 (a) $d \to \dfrac{d+10}{5}$ (b) $s \to 3(s+4)$

(c) $g \to \dfrac{g+4}{2}$ (d) $a \to \dfrac{a-10}{2}$

(e) $e \to 6e-2$ (f) $h \to \dfrac{h}{4}-2$

The pupil's explanations

Ⓒ Evaluating expressions (p 212)

C1 16

C2 (a) 8   (b) 20   (c) 60   (d) 0

C3 (a) $5 \to \mathbf{1}$ (b) $6 \to \mathbf{1\frac{1}{2}}$ (c) $10 \to \mathbf{3\frac{1}{2}}$

C4 (a) $1 \to 4$

(b) $1 \to 4$ and $2 \to 9$

(c) $1 \to 4$ and $5 \to 12$

(d) $1 \to 4$ and $5 \to 0$

(e) $2 \to 9$

(f) $4 \to 5$

C5 The pupil's three rules for $4 \to 0$

Cover up

One solution to board A:

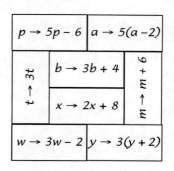

A solution to
board B:

$w \to 3w - 2$		$p \to 5p - 6$	$b \to 3b + 4$
$y \to 3(y + 2)$	$a \to 5(a - 2)$		
		$m \to m + 6$	
$t \to 3t$		$x \to 2x + 8$	

D Same but different! (p 213)

D1 (a) $+4$ (b) -6 (c) $+6$ (d) -4

E Equivalent expressions (p 214)

E1 (a) $2a + 8 = \mathbf{2}(a + 4)$

(b) $4a - 12 = 4(a - \mathbf{3})$

(c) $\frac{a-6}{2} = \frac{a}{2} - \mathbf{3}$

(d) $5a - 20 = \mathbf{5}(a - \mathbf{4})$

(e) $\frac{a+\mathbf{36}}{3} = \frac{a}{\mathbf{3}} + 12$

(f) $\mathbf{3}a + 6 = 3(a + \mathbf{2})$

E2 $3x + 18$ and $3(x + 6)$

$3(x - 2)$ and $3x - 6$

$3x - 18$ and $3(x - 6)$

The odd one is $3x - 2$.

E3 (a) $3x - 27$ (b) $\frac{b}{2} + 5$

(c) $5c - 150$ (d) $4d - 8$

(e) $0.5e + 3$ (f) $8w + 4$

(g) $\frac{a}{5} + 4$ (h) $70m - 7$

(i) $12 + 6f$ (j) $27 + 3y$

(k) $3 + \frac{k}{2}$ (l) $1.5n + 30$

E4 (a) $a \to 4(3a - 2)$, $a \to 12a - 8$;
28 when $a = 3$, 52 when $a = 5$

(b) $a \to 3(2a - 4)$, $a \to 6a - 12$;
6 when $a = 3$, 18 when $a = 5$

(c) $a \to 2(4a - 3)$, $a \to 8a - 6$;
18 when $a = 3$, 34 when $a = 5$

E5 (a), (b) Working leading to $6a + 10$

E6 (a) $8x + 2$ (b) $12 + 20y$

(c) $6z - 9$ (d) $30 + 15w$

(e) $16 - 4u$ (f) $v + 6$

(g) $4 + 5t$ (h) $3s - 4$

(i) $12 + 1.5r$ (j) $3 - \frac{1}{3}q$

E7 (a) $3(2a + 3) = \mathbf{6}a + 9$

(b) $4(4c + \mathbf{3}) = \mathbf{16}c + 12$

(c) $5(\mathbf{3}e + \mathbf{6}) = 15e + 30$

(d) $\mathbf{3}(2d + 7) = 6d + 21$

(e) $\frac{12e + \mathbf{8}}{\mathbf{4}} = 3e + 2$

(f) $\frac{8g + 24}{8} = g + \mathbf{3}$

F Inventing puzzles (p 216)

F1 You always get 5.

Think of a number.	n
Multiply it by 4.	$4n$
Add on 20.	$4n + 20$
Divide by 4.	$n + 5$
Take off your first number.	5
What is your answer?	5

F2 You get the number you first thought of.

Think of a number.	n
Multiply it by 2.	$2n$
Add on 10.	$2n + 10$
Divide by 2.	$n + 5$
Take off 5.	n
What is your answer?	n

F3 (a) You always get 0.

Think of a number.	n
Add on 5.	$n + 5$
Multiply it by 4.	$4n + 20$
Subtract 20.	$4n$
Divide by 4.	n
Take off your first number.	0
What is your answer?	0

(b) You always get 2.

Think of a number.	n
Add on 3.	$n + 3$
Multiply it by 2.	$2n + 6$
Subtract 6.	$2n$
Divide by 2.	n
Add 2.	$n + 2$
Take off your first number.	2
What is your answer?	2

F4 (a) Think of a number. n
Subtract 2. $n - 2$
Multiply it by 3. $3n - 6$
Add 14. $3n + 8$
Add your first number. $4n + 8$
Divide by 4. $n + 2$
Take off your first number. 2
What is your answer?   2

    (b) Think of a number. n
Subtract 4. $n - 4$
Multiply it by 5. $5n - 20$
Add 20. $5n$
Take off your first number. $4n$
Divide by 4. n
What is your answer? n

F5 The missing lines could be

        Multiply it by 2.
        Add on 18.
        Divide by 2.
        Take off your first number.

F6 ? is 16.

F7 Think of a number. n
    Multiply it by 3. $3n$
    Add 15. $3n + 15$
    Divide by 3. $n + 5$
    Take off 5. n

F8 The pupil's puzzle yielding 10

F9 The pupil's puzzle ending with start number

F10 (a) The pupil's input and outputs
     (b) The outputs differ by 3.
     (c) If the input number is n, we have

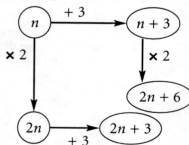

     (d) The pupil's further investigations

What progress have you made? (p 218)

1 (a) $y \xrightarrow{\times 3} 3y \xrightarrow{-6} 3y - 6$

   (b) $x \xrightarrow{+4} x + 4 \xrightarrow{\div 2} \dfrac{x + 4}{2}$

2 (a) 19   (b) $12\frac{1}{2}$ (c) $1\frac{1}{2}$

3 (a) $4x - 12$ (b) $5(3s + 2)$
   (c) $\dfrac{h}{2} + 6$ (d) $6y - 15$
   (e) $30z + 6$ (f) $4e + 2$

4 Think of a number. n
   Multiply it by 5. $5n$
   Add 50. $5n + 50$
   Divide by 5. $n + 10$
   Take off 3. $n + 7$
   Take off your first number. 7
   What is your answer?   7

Practice booklet

Sections A and B (p 71)

1 $1 \rightarrow$ **4**
     $4 \rightarrow$ **10**
     6 $\rightarrow 14$
     10 $\rightarrow 22$

2 $2 \rightarrow$ **14**
     $6 \rightarrow$ **38**
     3 $\rightarrow 20$
     0 $\rightarrow 2$

3 (a) $a \xrightarrow{-3} a - 3 \xrightarrow{\times 4} 4(a - 3)$
       $a \rightarrow 4(a - 3)$

   (b) $a \xrightarrow{\times 4} 4a \xrightarrow{-3} 4a - 3$
       $a \rightarrow 4a - 3$

   (c) $e \xrightarrow{\div 2} \dfrac{e}{2} \xrightarrow{-1} \dfrac{e}{2} - 1$
       $e \rightarrow \dfrac{e}{2} - 1$

   (d) $e \xrightarrow{-1} e - 1 \xrightarrow{\div 2} \dfrac{e - 1}{2}$
       $e \rightarrow \dfrac{e - 1}{2}$

(e)

$g \xrightarrow{\times 3} 3g \xrightarrow{-6} 3g - 6$

$g \rightarrow 3g - 6$

(f)

$h \xrightarrow{+8} h + 8 \xrightarrow{\div 4} \dfrac{h + 8}{4}$

$h \rightarrow \dfrac{h + 8}{4}$

(g)

$i \xrightarrow{-12} i - 12 \xrightarrow{\times 7} 7(i - 12)$

$i \rightarrow 7(i - 12)$

(h)

$s \xrightarrow{\times 8} 8s \xrightarrow{+2} 8s + 2$

$s \rightarrow 8s + 2$

4 (a)

$p \xrightarrow{\times 2} 2p \xrightarrow{+3} 2p + 3$

(b)

$m \xrightarrow{\times 2} 2m \xrightarrow{+6} 2m + 6$

(c)

$x \xrightarrow{+3} x + 3 \xrightarrow{\times 2} 2(x + 3)$

(d)

$b \xrightarrow{\div 3} \dfrac{b}{3} \xrightarrow{+10} \dfrac{b}{3} + 10$

(e)

$t \xrightarrow{+4} t + 4 \xrightarrow{\div 3} \dfrac{t + 4}{3}$

(f)

$n \xrightarrow{+1} n + 1 \xrightarrow{\times 5} 5(n + 1)$

(g)

$s \xrightarrow{\times 2} 2s \xrightarrow{+1} 2s + 1 \xrightarrow{\times 10} 10(2s + 1)$

(h)

$c \xrightarrow{\times 4} 4c \xrightarrow{+3} 4c + 3 \xrightarrow{\times 5} 5(4c + 3)$

or $5(3 + 4c)$

Sections C and D (p 72)

1 (a) (i) 10   (ii) 22   (iii) 2

   (b) (i) 35   (ii) $^-1$ (iii) 119

   (c) (i) 2   (ii) 1   (iii) 6.5

2 $3(m + 1)$ and $5(m + 1)$

3 (a) 4   (b) 1   (c) $^-5$

   (d) 3.5   (e) 30   (f) $^-4$

4 (a) $+1$ (b) -4 (c) $+4$ (d) $+2$

Sections E and F (p 73)

1 (a) $2(x + 3) = 2x + \mathbf{6}$

   (b) $\mathbf{4}(x + 4) = \mathbf{4}x + 16$

   (c) $5x + 10 = \mathbf{5}(x + \mathbf{2})$

   (d) $\dfrac{x - \mathbf{8}}{2} = \dfrac{x}{\mathbf{2}} - 4$

   (e) $\dfrac{x + 12}{\mathbf{2}} = \dfrac{x}{2} + \mathbf{6}$

   (f) $\mathbf{4}(x + 6) = \mathbf{4}x + 24$

2 (a) $2a + 6$ (b) $5x - 5$

   (c) $\dfrac{p}{4} + 2$ (d) $10p + 40$

   (e) $4s + 10$ (f) $100c - 100$

   (g) $\dfrac{y}{2} - 1.5$ (h) $30m - 6$

   (i) $8a - 4$ (j) $2j + 1$

   (k) $2k - 6$ (l) $40a - 10$

3 (a) The answer is always 2.

Think of a number.	x
Add on 3.	$x + 3$
Multiply it by 2.	$2x + 6$
Subtract 2.	$2x + 4$
Divide by 2.	$x + 2$
Subtract your first number.	2

   (b) The answer is always double the starting number.

Think of a number.	x
Multiply it by 5.	$5x$
Add on 10.	$5x + 10$
Divide by 5.	$x + 2$
Double it.	$2x + 4$
Subtract 4.	$2x$

   (c) The answer is the starting number.

Think of a number.	x
Subtract 1.	$x - 1$
Multiply by 3.	$3x - 3$
Add 6.	$3x + 3$
Divide by 3.	$x + 1$
Subtract 1.	x

   (d) The answer is 6 times the starting number.

Think of a number.	x
Multiply by 10.	$10x$
Subtract 5.	$10x - 5$
Divide by 5.	$2x - 1$
Multiply by 3.	$6x - 3$
Add 3.	$6x$

28 Decimals 2

p 219 **A** Multiplying and dividing by 10, 100, 0.1, …

p 220 **B** From 3 × 2 to 30 × 200, 0.3 × 0.2, …

p 222 **C** Rounding to one significant figure

p 223 **D** Estimation   Estimating by rounding to one significant figure

p 224 **E** Written multiplication

p 225 **F** Dividing by a decimal

p 226 **G** Written division

p 227 **H** Rounding to two or more significant figures

p 227 **I** Different ways of rounding

p 228 **J** Mixed questions

> **Practice booklet** pages 74 to 76

Ⓐ Multiplying and dividing by 10, 100, 0.1, … (p 219)

Ⓑ From 3 × 2 to 30 × 200, 0.3 × 0.2, … (p 220)

Pupils use tables facts and multiplication and division by 10 to multiply without a calculator.

◊ Each line in the calculation follows from the previous one by multiplying or dividing one of the numbers by 10. After some practice, pupils may not need to use so many steps. For example, 300 × 0.004 could be done by multiplying 12 by 10 twice and then dividing by 10 three times. The result is the same as if it had been divided by 10 once, that is 1.2 .

Ⓒ Rounding to one significant figure (p 222)

◊ Although we use the phrase 'significant figure', it would be better if it were 'significant place value'. The most significant place value is the highest one which is not occupied by a zero.

You can introduce the symbol ≈ to mean 'is approximately equal to' at this stage if you wish.

Ⓓ **Estimation** (p 223)

Pupils estimate the result of a calculation by rounding numbers to 1 s.f.

◊ Estimating the result of a division is more subtle than multiplication. For example, when estimating 476 ÷ 5.9 it is easier to think of 480 ÷ 6 than 500 ÷ 6. The division calculations in the exercise have been chosen to avoid this kind of subtlety.

Ⓔ **Written multiplication** (p 224)

◊ The position of the decimal point can be decided either by reference to the rough estimate or by the method used in section B, for example:
234 × 6 = 1404 leads to 23.4 × 6 = 140.4, and so to 23.4 × 0.6 = 1.404

Ⓕ **Dividing by a decimal** (p 225)

Ⓖ **Written division** (p 226)

◊ It helps to write the numbers with decimal points aligned, for example:
28.8
<u>0.24</u>

Ⓗ **Rounding to two or more significant figures** (p 227)

◊ Some pupils may think that in a number like 306 584, the second significant figure is the 6. This is an understandable mistake, seeing as the 0 signifies nothing! However, once the first significant place value has been identified, the next place value is the second, the next the third and so on.

H1 Parts (h) to (j) bring out the point above.

Another possible cause of confusion is when zeros are significant and must be recorded. For example, 2.976 to 3 s.f. is 3.00. This point is avoided in the questions. Depending on how confident the pupils feel, you could introduce it. But it might be better left for later or until it arises in the course of a particular problem.

I Different ways of rounding (p 227)

Pupils round numbers to a given number of significant figures or a given number of decimal places.

◊ At this stage it is probably too difficult for pupils to appreciate why there are two ways of rounding. It has to do with relative accuracy.
If a number is expressed to, say, 2 s.f., then whatever the size of the number, the error (from the third significant figure onwards) has about the same relative size. But if two numbers are both expressed to, say 2 d.p., the errors can be of very different relative sizes. The possible error in 352.68 (i.e. 0.005) is small relative to the size of the number, but in 0.68 it is much larger.

In measurements, a result expressed to, say, 3 s.f., has the same degree of accuracy whatever the metric units used. For example, the lengths 1.64 m, 164 cm and 1640 mm are expressed to the same degree of accuracy.

J Mixed questions (p 228)

A Multiplying and dividing by 10, 100, 0.1, ... (p 219)

A1 (a) 6.5   (b) 0.3678   (c) 0.2743
    (d) 12   (e) 0.84   (f) 0.0337
    (g) 5.46   (h) 0.003 08 (i) 501
    (j) 0.000 898 (k) 0.0262   (l) 2.37

A2 (a) 2300 m   (b) 37 g
    (c) 0.0173 km   (d) 2.08 litres
    (e) 0.009 77 m   (f) 0.023 04 km

A3 (a) 0.0865   (b) 367.8   (c) 2.243
    (d) 0.0569   (e) 0.0064   (f) 287
    (g) 0.536   (h) 70 800

A4 (a) 0.0023 m (b) 7.5 g   (c) 320 cm
    (d) 50.3 ml (e) 0.0305 m (f) 95.5 mm

A5 (a) 0.01   (b) 100   (c) 0.1
    (d) 1000   (e) 0.01   (f) 1000

A6 $23\,750 \times 0.001 = 23.75$
    $237.5 \times 0.01 = 2.375$
    $0.002\,375 \div 0.1 = 0.023\,75$

B From 3 × 2 to 30 × 200, 0.3 × 0.2, ... (p 220)

B1 (a) 6000   (b) 60 000   (c) 6
    (d) 60   (e) 0.06   (f) 2000
    (g) 2 000 000 (h) 200   (i) 2
    (j) 20

B2 (a) 0.18   (b) 0.12   (c) 180
    (d) 30   (e) 1.6

B3 (a) 5 × 6, 0.5 × 60, 50 × 0.6, 0.05 × 600, 500 × 0.06
    (b) 5 × 0.6, 0.5 × 6, 50 × 0.06, 0.05 × 60
    (c) 5 × 60, 0.5 × 600, 500 × 0.6, 50 × 6
    (d) 0.5 × 50, 0.05 × 500
    (e) 0.05 × 50, 0.5 × 5
    (f) 0.06 × 6

B4 (a) 32 m   (b) 500   (c) 50
    (d) 500   (e) 120

B5 (a) 6   (b) 1.6
    (c) 23.2 (− 10, × 0.4, × 0.2, + 20 or − 10, × 0.2, × 0.4, + 20)

B6 41

B7 $50 \times 40 = 2000$, $0.5 \times 400 = 200$,
$0.4 \times 50 = 20$

B8 $0.3 \times 4 = 1.2$, $0.2 \times 0.3 = 0.06$,
$30 \times 0.2 = 6$, $0.4 \times 30 = 12$

Ⓒ Rounding to one significant figure (p 222)

C1 The figure 4 is the most significant. 4000

C2 (a) 4000   (b) 8000   (c) 20 000
    (d) 800   (e) 800 000   (f) 40 000

C3 (a) 0.3   (b) 0.05
    (c) 0.003   (d) 0.0006

C4 (a) 0.07   (b) 100 000
    (c) 90 000   (d) 0.007

C5 (a) 50   (b) 2
    (c) 0.0008   (d) 0.08

Ⓓ Estimation (p 223)

D1 (a) $30 \times 80 = 2400$
   (b) $400 \times 30 = 12\,000$
   (c) $400 \times 200 = 80\,000$
   (d) $200 \times 4000 = 800\,000$
   (e) $5000 \times 400 = 2\,000\,000$
   (f) $2000 \times 300 = 600\,000$

D2 (a) $70 \times 9 = 630$ (b) $0.5 \times 2 = 1$
   (c) $0.02 \times 50 = 1$ (d) $0.8 \times 0.2 = 0.16$
   (e) $6 \times 0.02 = 0.12$

D3 (a) 1.0404   (b) 14.8732
   (c) 0.096 901   (d) 0.309 491 6

D4 $3 \times 10 \times 5000 = 150\,000\,\text{cm}$

D5 $20 \times 7 \times 3 = 420\,\text{m}^3$

D6 (a) $\dfrac{800 \times 0.1}{2} = 40$ (b) $\dfrac{0.5 \times 60}{10} = 3$

   (c) $\dfrac{0.08 \times 20}{4} = 0.4$ (d) $\dfrac{200}{4 \times 5} = 10$

(e) $\dfrac{6000}{60 \times 0.5} = 200$ (f) $\dfrac{900}{0.9 \times 50} = 20$

Ⓔ Written multiplication (p 224)

E1 (a) 33.6   (b) 1.89   (c) 20.8
   (d) 0.078   (e) 0.552   (f) 440
   (g) 0.0104   (h) 2.349

E2 (a) 12.48   (b) 12.48   (c) 124.8
   (d) 0.1248

E3 (a) 53.08   (b) 0.5308   (c) 53.08
   (d) 0.5308

E4 (a) 16.02   (b) 0.1602   (c) 16.02
   (d) 0.1602

E5 (a) 828
   (b) (i) 0.828   (ii) 8.28   (iii) 0.0828

E6 (a) 3.024   (b) 0.448   (c) 0.0855
   (d) 180.2

Ⓕ Dividing by a decimal (p 225)

F1 (a) 20   (b) 60   (c) 60
   (d) 30   (e) 8

F2 (a) 7   (b) 0.4   (c) 0.6
   (d) 0.2   (e) 300

F3 (a) 40   (b) 70   (c) 200   (d) 40
   (e) 200   (f) 2   (g) 80   (h) 5
   (i) 50   (j) 400

F4 40 m

F5 14 m

F6 15

F7 $0.004 + 0.02 = 0.024$, $0.08 \times 3 = 0.24$,
$1.6 \div 0.04 = 40$, $0.5 - 0.02 = 0.48$

Ⓖ Written division (p 226)

G1 (a) 0.805   (b) 39.6   (c) 44.8
   (d) 0.0072   (e) 160   (f) 735
   (g) 56   (h) 0.104   (i) 0.675
   (j) 855

G2 (a) 6.03   (b) 603   (c) 0.603
(d) 0.000603 (e) 6030

G3 (a) 346   (b) 346   (c) 0.003 46
(d) 3460   (e) 34.6

G4 (a) 6.025   (b) 6.025
(c) 0.006 025   (d) 0.060 25
(e) 602 000

G5 (a) 3.8   (b) 380   (c) 0.038
(d) 380

G6 (a) 16.5
(b) (i) 0.165   (ii) 16.5   (iii) 16.5

G7 (a) 185   (b) 1.425   (c) 15.5
(d) 0.1325

⊞ Rounding to two or more significant figures (p 227)

H1 (a) 67 000   (b) 0.067   (c) 0.15
(d) 460   (e) 0.0037   (f) 0.73
(g) 79   (h) 900   (i) 51 000
(j) 1000

H2 (a) 78 300   (b) 0.173   (c) 2000
(d) 3710   (e) 0.005 57   (f) 0.001 54
(g) 1280   (h) 903   (i) 851
(j) 49 900

H3 (a) 600 (b) 570 (c) 575 (d) 574.6

⏽ Different ways of rounding (p 227)

I1 (a) 5.33   (b) 5.3

I2 (a) 3600   (b) 4000

I3 (a) 6.6   (b) 6.52   (c) 0.008
(d) 0.009   (e) 6.5   (f) 0.007 82

I4 (a) 3460   (b) 3000   (c) 46 700
(d) 47 000   (e) 7.44   (f) 7.4

I5 (a) 20   (b) 20.04   (c) 0.006
(d) 0.005 89   (e) 5.098   (f) 5.10

⏽ Mixed questions (p 228)

J1 (a) 0.03   (b) 0.083   (c) 0.168
(d) 1.95   (e) 1.989   (f) 2.094

J2 (a) 30.23, 0.291   (b) 1503, 1.48
(c) 0.815, 7.85   (d) 1503, 30.23
(e) 30.23, 1.48

J3 (a) 1680   (b) 48   (c) 16.8
(d) 4.8   (e) 0.35   (f) 0.0168
(g) 350   (h) 0.48

J4 (a) 30   (b) 28.0   (c) 28.0
(d) 27.995   (e) 27.99

J5 (a) The result gets closer and closer to 0.25.
(b) The same happens.

What progress have you made? (p 229)

1 (a) 12   (b) 0.08   (c) 28

2 (a) 80 000   (b) 0.027   (c) 380
(d) 7.40

3 (a) $70 \times 0.5 = 35$ (b) $0.04 \times 300 = 12$
(c) $\dfrac{800 \times 0.2}{40} = 4$

4 (a) 2.482   (b) 0.0952

5 (a) 300   (b) 90

6 (a) 5.6   (b) 27

Practice booklet

Section A (p 74)

1 (a) 0.052   (b) 5.2
(c) 0.0634   (d) 634

2 (a) 100   (b) 0.01   (c) 1000
(d) 0.01   (e) 0.1   (f) 0.01

Section B (p 74)

1 (a) 15   (b) 1500   (c) 15
    (d) 150

2 (a) 2800   (b) 16   (c) 360000
    (d) 600   (e) 0.2   (f) 180

3 (a) 300×0.4; 30×4; 3×40; 0.3×400
    (b) 300×0.3; 30×3
    (c) 30×0.04; 3×0.4; 0.3×4; 0.03×40
    (d) 400×0.04; 40×0.4
    (e) 3×0.04; 0.3×0.4; 0.03×4

Sections C and D (p 74)

1 (a) 0.5   (b) 7000   (c) 20
    (d) 50000   (e) 9   (f) 0.7
    (g) 0.02   (h) 0.0008

2 (a) is given
    (b) $300 \times 50 = 15000$
    (c) $600 \times 300 = 180000$
    (d) $5000 \times 400 = 2000000$
    (e) $50 \times 10 = 500$
    (f) $0.3 \times 0.4 = 0.12$
    (g) $8 \times 90 = 720$
    (h) $6000 \times 0.8 = 4800$
    (i) $400 \times 0.008 = 3.2$
    (j) $0.03 \times 30 = 0.9$

3 3000×5 litres $= 15000$ litres

4 $40 \times 80p = 3200p = £32$

Section E (p 75)

1 (a) 16.17   (b) 0.111
    (c) 180   (d) 0.0228

2 (a) 4.76   (b) 47.6
    (c) 0.0476   (d) 476

Sections F and G (p 75)

1 (a) 7   (b) 8   (c) 90
    (d) 3000   (e) 300

2 (a) 2.4   (b) 4800   (c) 0.0148
    (d) 1.28   (e) 2500

Section H (p 76)

1 (a) 48000   (b) 0.042
    (c) 9.1   (d) 0.0037

2 (a) 871   (b) 0.0419
    (c) 27.8   (d) 310

3 (a) 0.00481   (b) 9.0
    (c) 4306000   (d) 421

Section I (p 76)

1 (a) 67800   (b) 68000
    (c) 5.4   (d) 5

2 (a) 0.04   (b) 0.040
    (c) 8710   (d) 8712.5

*3 (a) 6683, 6731, 6702
    (b) 6650   (c) 6749

*4 (a) £27800, £28428
    (b) £27500   (c) £28499.99

29 Investigations

Investigative and problem-solving work are best integrated into the development of mathematical concepts and skills. However, focusing on investigative work, as here, allows important skills of report writing to be developed. It is not intended that all the investigations should be done together or in the order given.

Optional
Counters, tiles or pieces of paper (for B1)
Square dotty paper (for B5)

Ⓐ Crossing points (p 230)

Discussion of the report is intended to highlight some important processes, for example, specialising, tabulating, generalising, predicting, checking, explaining, drawing conclusions. It is not intended to suggest there is one 'correct' way to approach an investigation and write up the findings.

◊ If pupils find it difficult to follow Chris's written work, try to involve them actively. They could read through the first half of page 213 (the 1, 2 and 3 line results) and then try with 4 lines.

Emphasise that the lines should be drawn as long as possible so that all crossings are shown.

Ask pupils to find as many different numbers of crossings as possible with 4 lines; 0, 1, 3, 4, 5 and 6 are possible:

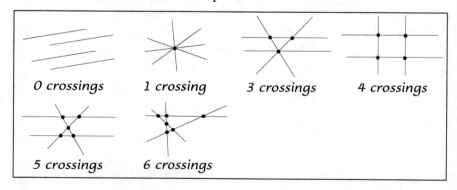

| 0 crossings | 1 crossing | 3 crossings | 4 crossings |

| 5 crossings | 6 crossings |

Ask how we can be sure that 2 crossings cannot be achieved or that 6 is the maximum number. Emphasise that pupils should try to explain their findings wherever possible.

◊ Now discuss how the investigation could proceed. Looking at the maximum number of crossing points is one choice (and it's the one made by the pupil in the write-up). Look at the results in the table. Pupils could try to spot a pattern before turning to page 233.

◊ Look at the table on page 233 and ask pupils if they can explain why the numbers of crossing points go up in the way they do. This is easier when pupils have found these results out for themselves. To get the maximum number of crossing points each additional line needs to cross all the lines in the diagram (except itself). Chris does not try to explain this in her report but without it she cannot be sure that the sequence of numbers continues in the way she describes. Point out that 'predict and check' may help to confirm results but is not foolproof (suppose there were actually 16 crossing points for 6 lines and your 'pattern' stopped you looking any further than 15).

◊ You could ask what the maximum number of crossing points would be with, say, 100 lines. Using Chris's method, this would take some time. With n lines, each line crosses $n-1$ lines to produce $n-1$ crossing points. However, $n(n-1)$ counts each crossing point twice so the number of crossing points is $\frac{n(n-1)}{2}$.

◊ As a further investigation, pupils can look at the maximum number of closed regions obtained. An interesting result is that you get the same set of numbers but with 0 included: 0, 1, 3, 6, 10, 15, …

Ⓑ Some ideas (p 234)

B1 Round table (p 234)

> Optional: Counters, tiles or pieces of paper may be useful to represent the people round the table.

◊ Pupils may find it helpful to do the investigation by moving counters or tiles (labelled A to E) round a drawing of a table. There is only one other arrangement:

Ask pupils to consider how they know there are no other arrangements. They may realise that every person has sat next to every other person so no other arrangements are possible. Ask pupils if that is what they expected – often pupils think there will be more possibilities.

◊ In one school, the investigation proved easier when the table was 'unrolled' into a straight line (remembering that the two end people are in fact sitting next to each other). For five people the two different arrangements are

ABCDE   ADBEC

◊ The results for 3 to 10 people are

Number of people	Number of arrangements
3	1
4	1
5	2
6	2
7	3
8	3
9	4
10	4

For an even number of people the rule is $a = \frac{p-2}{2}$
and for an odd number of people the rule is $a = \frac{p-1}{2}$
where a is the number of arrangements
and p is the number of people.

Since each person has two neighbours, the number of arrangements has to be the number of complete pairs of people that can sit next to an individual. These formulas give the number of complete pairs.

B2 Nine lines (p 234)

◊ Clarify that the grids of squares have to be drawn with straight lines either parallel to each other or at right angles.

As in 'Crossing points' lines should be drawn as long as possible so that all possible squares are counted. For example, diagrams such as this are not considered valid.

◊ With 9 lines 0, 6, 10 and 12 squares are possible:

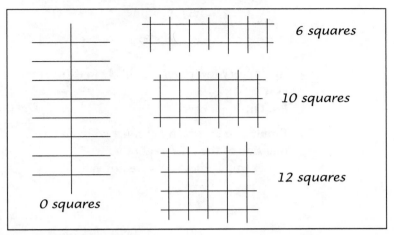

6 squares

10 squares

12 squares

0 squares

9 parallel lines will also produce 0 squares.

◊ All possible numbers of squares for 4 to 12 lines are

Number of lines	Numbers of squares				
4	0	1			
5	0	2			
6	0	3	4		
7	0	4	6		
8	0	5	8	9	
9	0	6	10	12	
10	0	7	12	15	16
11	0	8	14	18	20
12	0	9	16	21	24	25

Pupils who produce a full set of results as in the table may make various observations such as:

• The minimum is always 0 and this can be achieved by a set of parallel lines.

• The next possible number of squares is $n - 3$ for n lines. These numbers go up by 1 each time.

◊ Encourage pupils to follow their own lines of investigation. For example, they could consider the maximum number of squares possible each time.

Encourage pupils to describe how the lines should be arranged to give the maximum number of squares by asking questions such as 'How would you arrange 100 lines to achieve the maximum number of squares?' and 'What about 99 lines?'

Formulas are

even numbers of lines: $s = (\frac{n}{2} - 1)^2$

odd numbers of lines: $s = (\frac{n}{2} - \frac{1}{2})(\frac{n}{2} - \frac{3}{2})$

Again, very few pupils are likely to express their conclusions algebraically at this stage.

B3 Cutting a cake (p 235)

◊ Emphasise that each cut must go from one side of the cake to another.
For example, these cuts are not valid.

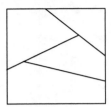

◊ Pupils could investigate the minimum and maximum number of pieces. The results are

Number of cuts	Minimum number of pieces	Maximum number of pieces
0	1	1
1	2	2
2	3	4
3	4	7
4	5	11
5	6	16
6	7	22
7	8	29
8	9	37

◊ Pupils may comment that:
  - The minimum number of pieces goes up by 1 each time.
  - The minimum number of pieces is always 1 more than the number of cuts (or, with n cuts, the minimum number of pieces is $n + 1$).
  - The minimum number of pieces can be achieved by making a set of parallel cuts.

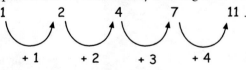

  - The maximum number of pieces goes up by 1, then 2, then 3 and so on.
  - To achieve the maximum number of pieces, take the previous diagram and make a cut that crosses each of the previous cuts.
  - The sequence of numbers for the maximum number of pieces for one or more cuts (2, 4, 7, 11, 16, 22, …) can be found by adding 1 to each of the numbers in the sequence for the maximum number of crossing points in 'Crossing points' (1, 3, 6, 10, 15, 21, …).

◊ Pupils may correctly predict the maximum number of pieces for various numbers of cuts but find it difficult to produce the corresponding diagrams. Using larger squares may help.

◊ For any number of cuts all numbers of pieces between the minimum and maximum can be achieved. Each diagram can be found by altering the previous one. For example, with four cuts:

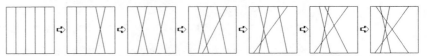

5 pieces   6 pieces   7 pieces   8 pieces   9 pieces   10 pieces   11 pieces

◊ With n cuts, the maximum number of pieces is $\frac{n(n-1)}{2} + 1$.

B4 Matchstick networks (p 235)

This investigation involves three variables.

◊ The 'standard' way to approach a three-variable problem is to keep one variable constant and investigate the relationship between the other two. However in this case the relationship between the three variables is easy enough for pupils to spot from a table, for example:

Enclosed spaces (E)	Nodes (N)	Matches (M)
0	8	7
1	8	8
2	9	10
2	7	8
4	6	4
3	8	10

The rule is $E + N - 1 = M$

◊ If the rule is not found in this way, suggest that pupils first investigate networks with no enclosed spaces ('trees'), to find the rule $N - 1 = M$. Then look at one enclosed space, two enclosed spaces, and so on.

◊ The explanation of the rule is quite difficult at this stage, but you can encourage pupils to explain what can happen when one more match is added to a network

M up by 1 N up by 1 E the same	M up by 1 N the same E up by 1	M up by 1 N the same E up by 1	M up by 1 N up by 1 E the same

◊ The full explanation is as follows.
For the simplest network ●—● $E = 0, N = 2, M = 1$
so $E + N - M = 1$ for this network.

Whenever a new match is added to the network, $E + N - M$ does not change (see the four cases above). So $E + N - M$ is always 1.

B5 Turn, turn, turn (p 236)

This is a very rich activity with many possibilities for extension. Encourage pupils to follow their own lines of enquiry but some may need help in formulating questions.

Optional: Square dotty paper

◊ Many have found it beneficial for pupils to walk through the instructions to draw a turning track, emphasising the 90° turn each time.

◊ Encourage pupils to investigate questions such as:

  • Do you always get back to your starting point?
    If you do, how many times do you repeat the instructions?

  • What difference does it make if you turn left instead of right each time?

  • What shape are the tracks?
    Do you get different types of tracks with different sets of numbers?

  • Why do some tracks have 'holes'
    while others have 'overlaps'?

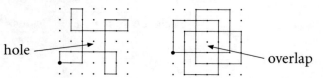

    Can you predict whether or not a track will have a hole or an overlap from the set of three numbers? Can you predict the size of the hole?

  • What will happen with sets of consecutive numbers?

  • What if you look at sets of numbers where the first two are always the same?

  • What happens if you change the order of the numbers?
    Will the track for 1, 2, 4 look like the track for 4, 1, 2 for example?

  • What if two or more of the numbers are the same?

◊ Pupils may notice these facts.

  • It doesn't matter what order the numbers are in. You always produce the same track although it may be rotated or reflected.

  • Turning left produces a reflection of the turning right track.

  • If the numbers are all the same, you get a square track.

  • If two of the numbers are the same, you get a cross shape.

  • All tracks made with three numbers have rotation symmetry.

- If the two smallest numbers add up to the largest, you get a track with no hole or overlap.
  For example,

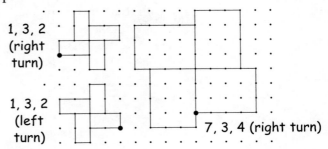

1, 3, 2 (right turn)

1, 3, 2 (left turn)

7, 3, 4 (right turn)

- If the sum of the two smallest numbers is greater than the largest number, you get a 'windmill' shape with an overlap.
  For example,

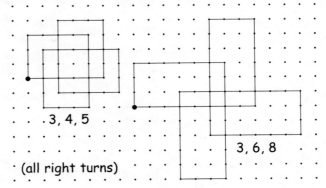

3, 4, 5

3, 6, 8

(all right turns)

- If the sum of the two smallest numbers is smaller than the largest number, you get a windmill shape with a hole.
  For example,

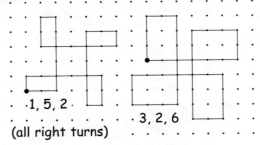

1, 5, 2

3, 2, 6

(all right turns)

The size of the hole is the largest number minus the sum of the two smallest numbers.

◊ Investigating longer sets of numbers, pupils will find that four numbers produce infinite 'spiral' tracks and five numbers produce 'closed' tracks with rotation symmetry.

For example,

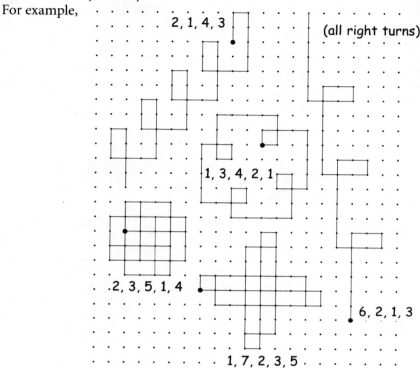

2, 1, 4, 3   (all right turns)

1, 3, 4, 2, 1

.2, 3, 5, 1, 4

6, 2, 1, 3

1, 7, 2, 3, 5

◊ Pupils can investigate what happens if you change the turning angle. For example, you could use a 60° right turn and investigate on triangular spotty paper.

◊ Pupils can use LOGO to draw their turning tracks.

30 Parallel lines

Essential

Set square

Practice booklet pages 77 to 79

A Using parallel lines (p 237)

Set square

◊ Pupils are likely to have met the word 'parallel' before. You could ask a pupil to draw a pair of parallel lines on the board and then ask the rest of the class to say what it is that makes the lines parallel.

◊ We are restricting our use of 'parallel' to straight lines, though the word is used in everyday speech to refer to curved lines (railway lines, lines of latitude) that never meet.

B Parallel lines and angles (p 240)

◊ The pencil idea can be used on an OHP transparency of the diagram for question B3.

Other acceptable ways to justify corresponding and alternate angles may come up in discussion. For example, you can think of angle x as a 'solid thing' sliding along a line to become a corresponding angle; corresponding angles together with vertically opposite angles can be used to justify alternate angles.

The fact that x and y add up to 180° (that is, are supplementary) can also emerge from the B3 diagram.

C Related angles (p 242)

◊ In one school, long rulers were placed on the floor like this.

Pupils were asked to stand on a vertically opposite/corresponding/alternate angle to the one given.

Ⓓ Explaining how you work out angles (p 244)

◊ It is important for pupils to be able to explain the steps of their reasoning, even though these may appear obvious.

It is generally easier, in cases where only angles are involved, to use the method on page 244. This may involve labelling angles other than those given, hence the need to show the new labels on a copy of the diagram.

Pupils may need help on page 246 with using letters for points and three letters to label angles.

Ⓐ Using parallel lines (p 237)

A1 They have a different slope: *a* goes 1 square up for every 2 along while *b* goes 1 square up for every 3 along.

A2 (a) Yes, they both go 2 squares down for every 3 along.

(b) No, *s* is steeper than *r*: *s* goes 4 down for every 2 along, but *r* goes less than 4 down for 2 along.

A3 *a, b, h* and *j*   *c, i* and *k*   *d, f* and *g*   *e* and *l*

A4
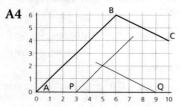
The diamond is at (5, 2).

A5
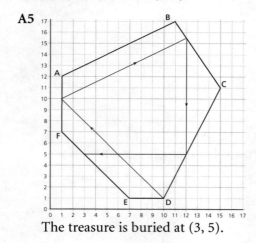
The treasure is buried at (3, 5).

A6 The two lengths measured on each transversal are the same.
If you mark points 12 cm apart and 4 cm apart, the lengths measured on each transversal are in the ratio 2 : 1.
In general, if the distances between the dots on the parallel lines are in the ratio *n* : 1, the lengths measured on each transversal are in the ratio *n* − 1 : 1.

Ⓑ Parallel lines and angles (p 240)

B1 They are parallel (they go in the same direction).

B2 They are parallel (they go in the same direction).

B3 The pupil's sketch

B4 (a) $a = 130°$, $b = 50°$, $c = 130°$, $d = 50°$

(b) $e = 110°$, $f = 70°$, $g = 110°$, $h = 70°$

(c) $i = 108°$, $j = 72°$, $k = 108°$, $l = 72°$

B5 $a = 42°$, $b = 138°$
$c = 85°$, $d = 95°$
$e = 37°$, $f = 143°$

Ⓒ Related angles (p 242)

C1 (a) Angles *a* and *b* are **corresponding** angles.

(b) Angles *c* and *d* are **alternate** angles.
Angles *c* and *e* are **corresponding** angles.
Angles *d* and *e* are **vertically opposite** angles.

(c) Angles *f* and *g* are **alternate** angles.
Angles *f* and *h* are **supplementary** angles.
Angles *g* and *h* are **supplementary** angles.
Angles *g* and *i* are **vertically opposite** angles.

C2 (a) 65° corresponding angles

(b) 55° alternate angles

(c) 112° vertically opposite angles

(d) 75° supplementary angles

(e) 106° supplementary angles

(f) 120° alternate angles

D Explaining how you work out angles (p 244)

Reasons need to be given for steps of working.

D1 (a) $x = 75°$

(b) $y = 95°$, $z = 65°$

(c) $v = 59°$, $w = 138°$

D2 (a) $x = 94°$, $v = 86°$

(b) $u = 76°$, $v = 99°$

(c) $r = 72°$

D3 (a) $x = 112°$, $y = 75°$

(b) $u = 132°$, $v = 56°$

(c) $p = 47°$, $q = 63°$

D4 (a) $x = 40°$ (b) $y = 110°$

(c) $z = 24°$ (d) $u = 70°$, $v = 60°$

(e) $w = 95°$ (f) $g = 105°$

D5

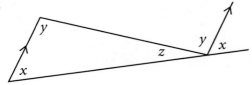

x : corresponding angles
y : alternate angles
$x + y + z = 180°$
(angles on a straight line)
The angles of a triangle add up to 180°.

D6

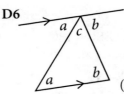

a : alternate angles
b : alternate angles
$a + c + b = 180°$
(angles on a straight line)

D7 (a) $p = 43°$, $q = 137°$

(b) $r = 84°$, $s = 142°$

(c) $t = 70°$, $u = 110°$, $v = 70°$

(d) $w = 38°$, $x = 104°$

(e) $a = 47°$, $b = 23°$, $c = 110°$

D8 Here are some possible explanations.

(a) Angle HGB + angle DGH = 180°
(angles on a straight line)
So **angle HGB = 63°**

Angle ADC = angle DGF (corresponding)
= 180° – 117° = **63°**

(b) Angle IJM = angle OMN = 63° (corresponding angles)

Angle IJL + angle IJM = 180° (angles on a straight line)
So **angle IJL = 117°**

(c) Angle UTQ = 180° – 109° = 71°

Angle PQS = angle UTQ = **71°** (corresponding angles)

D9 Lines AD and BC are parallel (angles DAC and ACB are alternate angles).

Lines DE and CF are parallel (angles EDF and DFC are alternate angles.)

Lines AC and EF are parallel (angles ACF and CFE are supplementary angles between parallels).

What progress have you made? (p 247)

1 The pupil's drawing

2 They are not parallel.
On the top line you go up 2 when you go along 3. On the other line you go up more than 2 when you go along 3.

3 $a = 60°$ $b = 95°$ $c = 35°$

4 Angle DEB + angle BEF = 180°
(supplementary angles on a straight line)
So **angle DEB = 80°**

Angle EFC = angle DEB = **80°**
(corresponding angles)

Angle BCF = angle GBC = **75°**
(alternate angles)

Practice booklet

Section A (p 77)

1 (a) The pupil's drawing
  (b) The pupil's drawing
  (c) The 'diamond' shape has its sides all
      the same length.
      It has two pairs of supplementary
      angles. The diagonals cut one
      another in half, crossing at right
      angles.

2 (3, 4)

3 (6, 4)

Sections C and D (p 78)

1 Angles a and d are **vertically opposite**.
  Angles e and a are **corresponding**.
  Angles c and f are **alternate**.
  Angles f and d are **supplementary**.
  Angles h and d are **corresponding**.
  Angles f and e are **supplementary**.
  Angles f and g are **vertically opposite**.
  Angles b and f are **corresponding**.

2 (a) $p = 115°$, $q = 108°$
  (b) $r = 50°$, $s = 45°$
  (c) $t = 48°$, $u = 75°$
  (d) $v = 62°$, $w = 68°$

3 (a) $r = 29°$, $s = 151°$, $t = 47°$, $u = 47°$
  (b) $v = 35°$, $w = 145°$, $x = 37°$, $y = 143°$

4 (a) $p = 63°$, $q = 45°$
  (b) $r = 70°$, $s = 62°$
  (c) $t = 62°$, $u = 30°$
  (d) $v = 117°$
  (e) $w = 38°$
  (f) $x = 46°$

5 (a) Angle EDG = 100° (Angles EDG and
      CDG are supplementary angles on a
      straight line.)

      Angle FEH = 100° (Angles EDG and
      FEH are corresponding angles.)

  (b) Angle LMJ = 110° (Angles OLM and
      LMJ are alternate angles.)

      Angle JMN = 70° (Angles LMJ and
      JMN are supplementary angles on a
      straight line.)

  (c) Angle RSX = 85° (Angles VST and
      RSX are vertically opposite angles.)

      Angle QRW = 85° (Angles RSX and
      QRW are corresponding angles.)

Percentage

This unit covers changing a percentage to a decimal, calculating a percentage of a quantity and expressing one quantity as a percentage of another. The latter leads on to drawing pie charts using a circular percentage scale.

Essential

Pie chart scale

Practice booklet pages 80 to 84

A Understanding percentages (p 248)

◊ Proportions are given in a variety of ways. From class or group discussion should emerge the need for a 'common currency' for expressing and comparing proportions.

◊ French cheeses may be labelled '40% matière grasse', for example; this means 40% of the dry matter is fat (i.e. the water content of the cheese is discounted).

◊ The order is:
Mascarpone 45%, Blue Stilton 36%, Red Leicester $33\frac{1}{3}$%, Danish Blue 28%, Edam 25%, Camembert 21%, Cottage Cheese 2%, Quark 0.2%

B Percentages in your head (p 250)

Mental work with percentages can be returned to frequently in oral sessions.

C Percentages and decimals (p 251)

D Calculating a percentage of a quantity (p 252)

It is important that pupils should feel confident about their method of working out a percentage of a quantity. The 'one-step' method is more sophisticated but ultimately better because it easily extends to a succession of percentage changes.

E Changing fractions to decimals (p 253)

You could point out that the division symbol ÷ is itself a fraction line with blanks above and below for numbers.

F One number as a percentage of another (p 254)

The approach used here depends on conversion from decimal to percentage.

G Drawing pie charts (p 256)

> Pie chart scale

◊ The pie charts shown are not labelled with the percentages, in order to give practice in measuring. However, it is a good practice to include the percentages.

◊ Rounding often leads to percentages which add up to slightly more or less than 100%. For this reason, when drawing pie charts it is often better to work to the nearest 0.1% as the small excess or deficit can be safely ignored.

H Problems involving percentages (p 258)

A Understanding percentages (p 248)

A1 (a) About 27%   (b) Water
    (c) About 21%

A2 (a) C, D, F   (b) G, H
    (c) A, E   (d) B, I

A3 (a) 20%–30%   (b) 85%–95%
    (c) 45%–55%   (d) 55%–65%
    (e) 3%–8%

B Percentages in your head (p 250)

B1 (a) $\frac{1}{4}$ (b) $\frac{3}{4}$ (c) $\frac{1}{10}$
    (d) $\frac{9}{10}$ (e) $\frac{1}{5}$

B2 (a) 50%   (b) 10%   (c) 25%
    (d) 75%   (e) $33\frac{1}{3}$%, 25%

B3 (a) £15   (b) £42   (c) £17.50

B4 (a) £10   (b) £21   (c) £17.50

B5 (a) To find 10%, you can divide by 10.
    (b) To find 5%, you can divide by 10, then halve.

B6 (a) 1p   (b) 3p   (c) 37p

B7 (a) £10.50   (b) £38.50

C Percentages and decimals (p 251)

C1 (a) 0.5 (b) 0.25 (c) 0.65 (d) 0.78
    (e) 0.1 (f) 0.01 (g) 0.04 (h) 0.4

C2

$$\frac{45}{100} = \mathbf{0.45} = \mathbf{45\%}$$
$$\frac{\mathbf{57}}{\mathbf{100}} = 0.57 = \mathbf{57\%}$$
$$\frac{5}{100} = \mathbf{0.05} = \mathbf{5\%}$$
$$\frac{\mathbf{63}}{\mathbf{100}} = \mathbf{0.63} = 63\%$$
$$\frac{7}{\mathbf{100}} = 0.07 = \mathbf{7\%}$$

C3 (a) 30% (b) 80% (c) 83% (d) 3%

C4 1%, 0.1, $\frac{12}{100}$, 15%, 0.25, 0.3, $\frac{45}{100}$

D Calculating a percentage of a quantity (p 252)

D1 (a) 162 g   (b) 256.2 g (c) 66.7 g
    (d) 93.8 g   (e) 26.6 g (f) 35.2 g
    (g) 201.6 g (h) 52.8 g

D2 (a) 2.25 g   (b) 6.72 g
     (c) 3.6 g   (d) 1.68 g

D3 0.6 g

D4 He is right for 10%, but dividing by 5 gives 20%, not 5%.

D5 (a) Sugar 19.95 g, fat 10.5 g, protein 2.8 g
     (b) Sugar 85.5 g, fat 45 g, protein 12 g
     (c) Sugar 285 g, fat 150 g, protein 40 g

E Changing fractions to decimals (p 253)

E1 (a) 0.25   (b) 0.125   (c) 0.05
     (d) 0.8   (e) 0.375   (f) 0.875
     (g) 0.28   (h) 0.15   (i) 0.22
     (j) 0.9375

E2 $\frac{29}{50}$ (0.58), $\frac{3}{5}$ (0.6), $\frac{5}{8}$ (0.625), $\frac{13}{20}$ (0.65)

E3 (a) 0.14 (b) 0.57 (c) 0.11 (d) 0.56
     (e) 0.64 (f) 0.27 (g) 0.08 (h) 0.38
     (i) 0.41 (j) 0.87

F One number as a percentage of another (p 254)

F1 71% (to the nearest 1%)

F2 (a) 60%   (b) 35%   (c) 87.5%
     (d) about 33%   (e) about 67%

F3 (a) 29%   (b) 78%   (c) 23%
     (d) 41%   (e) 5%

F4 (a) $\frac{1}{4}$ (b) 25%   (c) 19%

F5 30%

F6 Tina's method is correct.

F7 (a) 16%
     (b) In a class of 30 it would be about 5 people.

F8 12.4 × 0.41 = 5.084

F9 (a) 64%   (b) 54%   (c) 76%
     (d) 35%   (e) 74%   (f) 69%

F10 A 14.0% fat   B 15.5% fat
B has higher percentage of fat.

G Drawing pie charts (p 256)

G1 These features (amongst others) may be noticed:
Cheese spread has higher proportions of water and carbohydrate, but lower proportions of fat and protein.

G2 (a) 27%   (b) 23%   (c) 46%   (d) 53%

G3

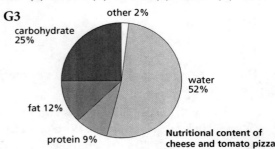

Nutritional content of cheese and tomato pizza

G4 (a) Meat, fish and eggs   (b) 9%
     (c) Clio is wrong. The chart shows money spent, not the amounts eaten.

G5

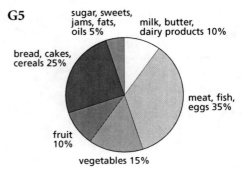

G6 (a) 34%
     (b) Foreign news 22%, sport 16%, entertainment 9%, finance 19%

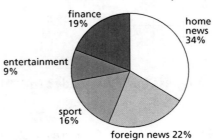

G7

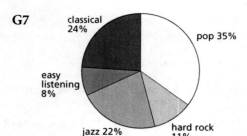

classical 24%
pop 35%
easy listening 8%
jazz 22%
hard rock 11%

Ⓗ **Problems involving percentages** (p 258)

H1 287.5 ml

H2 40 g

H3 (a) No. If he gets 30 g protein he will get 60 g fat.

(b) With cheese B, if he gets 30 g protein he will get 22.5 g fat. With cheese C, if he gets 30 g protein he will get 15 g fat. So cheese C will suit him.

H4 (a) 50%   (b) 37.5%
     (c) 91.7%   (d) 58.3%

H5 12.1%

H6 (a)

of	20	**10**	50
5%	**1**	0.5	**2.5**
1%	0.2	0.1	0.5
8%	**1.6**	0.8	**4**

(b)

of	50	**30**	80
15%	**7.5**	**4.5**	12
90%	45	27	**72**
1%	0.5	**0.3**	0.8

What progress have you made? (p 259)

1 (a) C   (b) A   (c) E

2 (a) 0.5 (b) 0.45 (c) 0.04 (d) 0.07

3 (a) £5   (b) 2 kg   (c) 9 kg   (d) £2

4 (a) 68.4 g   (b) 13 g

5 73%

6 21 out of 25 (84%) is better than 30 out of 37 (81%).

7

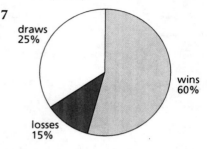

draws 25%
wins 60%
losses 15%

Practice booklet

Section A (p 80)

1 Cadbury's Double Decker: protein 6%, fat 22%, carbohydrate 72%

Sainsbury's Creamy White: protein 7%, fat 35%, carbohydrate 58%

2 (a) 30% (b) 10% (c) 70% (d) 50%

Section B (p 81)

1 (a) £10   (b) £8.50   (c) £45

2 (a) £15   (b) £13   (c) £12.50

3 (a) 7 kg   (b) £7.50   (c) £7.30

4 (a) 5p   (b) 15p   (c) £4.85
     (d) 90p   (e) £4.50

5 (a) Divide by 10 to find 10%.
     (b) Divide by 10, then halve to get 5%.
     (c) Add the results for 10% and 5% together.

6 (a) £1.50   (b) £6   (c) £13.50

7 (a) 3% of £10 is £0.30.
         All the others are £3.

(b) 75% of £8 is £6.
All the others are £5.

(c) 5% of £80 is £4.
All the others are £4.50.

Section C (p 82)

1 (a) 0.68   (b) 0.41   (c) 0.8
   (d) 0.08   (e) 0.09   (f) 0.9
   (g) 0.545   (h) 0.305

2 (a) 65%   (b) 60%   (c) 6%
   (d) 6.5%   (e) 2%   (f) 22%
   (g) 20%   (h) 20.2%

3 $\frac{58}{100}$ = **0.58** = **58%**

   $\frac{32}{100}$ = 0.32 = **32%**

   $\frac{9}{100}$ = **0.09** = **9%**

   $\frac{80}{100}$ = **0.8** = 80%

   $\frac{5}{100}$ = 0.05 = **5%**

4 (a) 7% $\frac{8}{100}$ 0.09   0.1   0.8 $\frac{81}{100}$
   (b) 0.02   0.1   12%   20% $\frac{1}{2}$ 0.7
   (c) 0.07 $\frac{8}{100}$ 0.3 and $\frac{3}{10}$ 40%   75%

Section D (p 82)

1 28% of 4 = 1.12
   47% of 220 = 103.4
   91% of 7.3 = 6.643
   8% of 8 = 0.64

   The answer for the extra calculation is
   26% of 82 = 21.32

2 A is the odd one out (£10).
   The others are £6.

3 A is the odd one out (3.2 kg).
   The others are 3.3 kg.

4 (a) £1199.25   (b) £6795.75

5 (a) £410.50   (b) £36 306   (c) £2856

Sections E and F (p 83)

1   $\frac{13}{27}$ (0.48), $\frac{6}{11}$ (0.55), $\frac{11}{19}$ (0.58)

2 76%

3 12%

4 (a) 31%   (b) 69%

5 B is the odd one out (24.7%). Others are 25%.

6 B is the odd one out (56%). Others are 60%.

7 (a) 8.7%   (b) 88.4%   (c) 2.9%

Sections G and H (p 84)

1 (a), (b) The percentages for the pie charts are

	Protein	Carbohydrate	Fat	Fibre
QO	13%	70%	9%	8%
AB	14%	50%	4%	32%

   (c) The protein content of each are similar,
        but All Bran contains a lot more fibre
        and less carbohydrate and fat than
        Quaker Oats.

2 (a) The percentages for the pie charts are

	Coal	Petroleum	Natural gas	Nuclear
1982	34.7%	36.2%	23.0%	6.1%
1995	22.9%	35.1%	32.2%	9.8%

   (b) The use of coal has dropped considerably.
        Natural gas and nuclear energy have
        increased their shares.

3 (a) 30p   (b) 13%

4 (a)

of	5	82	50
10%	0.5	**8.2**	**5**
20%	**1**	**16.4**	10
70%	3.5	**57.4**	35

   (b)

of	30	55	20
30%	**9**	**16.5**	6
80%	24	44	16
20%	**6**	11	**4**

32 Think of a number

This unit develops the important idea of an inverse process and shows how to use it to solve an equation.

In later work, pupils should realise the limitations of using arrow diagrams to solve equations. The 'balancing' method should then become the principal method for solving equations.

In this unit, arrow diagrams are drawn with circles and ellipses. Pupils may find it easier to use squares and rectangles.

T	p 260 **A** Number puzzles	Using arrow diagrams and inverses to solve 'think of a number' puzzles
T	p 262 **B** Using letters	Linking equations, arrow diagrams and puzzles
T	p 263 **C** Solving equations	Solving linear equations by using arrow diagrams and inverses
T	p 264 **D** Quick solve game	A game to consolidate solving equations

Essential	**Optional**
Calculators	Sheet 158
Sheet 157	
Practice booklet pages 85 to 87	

A Number puzzles (p 260)

Calculators for working with decimals and large numbers

T

'This introduction went down very well.'
'It was a valuable run up to solving equations.'

◊ As a possible introduction, ask the pupils each to think of a number without telling anyone what it is. Now ask them to:

Add 1.

Multiply by 3.

Add 5.

Take away 2.

Divide by 2.

Ask some pupils to tell you what their answers are and then work backwards to give the numbers they were thinking of.
Pupils could discuss how they think you worked them out.

Now give some single-operation problems such as

> 'I think of a number.
> I divide by 0.2 and my answer is 15.
> What number did I think of?'

Ensure discussion of these brings out the idea of using an inverse operation in a reversed arrow diagram.

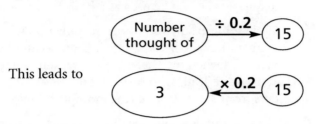

This leads to

Encourage pupils to check each solution by substituting in the original puzzle.

Now move on to 'think of a number' problems that use more than one operation and show how arrow diagrams and inverses can be used to solve them.

Some pupils will feel more confident using 'trial and improvement' to solve these problems. To demonstrate the power of using inverses, include problems involving decimals and many operations.

For example,

> 'I think of a number.
> I divide by 5.
> I subtract 45.
> I multiply by 1.2.
> I add 8.
> I multiply by 0.2.
> I subtract 12.
> My answer is 10.
> What number did I think of?'

Pupils could use both trial-and-improvement and inverse methods and compare them.

You could end this teacher-led session by asking the pupils to solve the puzzles on the pupil's page. The solutions to these puzzles are 18, 12 and 15.3 respectively.

A6 Pupils who are struggling with this can be reminded of work from 'Inputs and outputs'. Ask them to consider what happens if n is used as the 'input' for this puzzle.

'Some students were annoyed at going back to "Inputs and outputs", but I feel it was a good idea.'

B **Using letters** (p 262)

◊ It may help to tell pupils that this section is not about solving the puzzles or equations, but about linking equations, arrow diagrams and puzzles. Otherwise, they may feel that they have not fully answered the questions unless they have solved each puzzle to find the number thought of.

The questions could be used as a basis for a whole-class discussion.

◊ Include some examples in your introduction where multiplication is the first operation and some where it is the second operation. Pupils must understand when they need to use brackets. You could include the puzzle and diagrams leading to the equation $2n + 3 = 20$, bringing out how it differs from $2(n + 3) = 20$ (the example on page 262).

◊ Make sure pupils know that $2 \times (n + 3)$ is the same as $(n + 3) \times 2$ and that $2(n + 3)$ is shorthand for it.

◊ Introduce the 'division line' and make sure pupils are aware that, for example,
$(n \div 4) + 5 = 7$ can be written $\frac{n}{4} + 5 = 7$

and $(n + 5) \div 4 = 7$ can be written $\frac{n+5}{4} = 7$.

C **Solving equations** (p 263)

Calculators for working with decimals and large numbers

'Pupils made up their own equations for others to solve.'

◊ Pupils who have already used balancing ideas may want to extend the method here. Point out that it is possible to use the balancing approach in this section but it needs to be modified to deal with equations involving subtraction. However, explain that using arrow diagrams to solve equations should improve their understanding of the general processes involved and help with later work.

◊ In your discussion, include the equation $2n - 3 = 130$ and compare it with $2(n - 3) = 130$ on the pupil's page.

◊ The equations in question C5 involve more than two operations. It may be beneficial to include some examples of this kind in your introduction.

D **Quick solve game** (p 264)

'This was worthwhile. It had a different feel – pupils were handling it in their heads.'

The version of this game described in the pupil material can be played in groups of three or four. It can also be played as a whole class or individually (see below).

> Each group needs a set of 36 cards (Sheet 157)
> Optional: Each pupil needs a copy of sheet 158 if they check each other's answers.

◊ Emphasise that *all* players take a card at the start of the game and take another as soon as they think they have solved their equation. They should keep their equation cards – they will not be used by another player.

◊ There are other ways to use the cards.

One pile version

This is played as the version in the book but pupils take cards from a mixed shuffled pile.

Three pile whole-class version

The game could be played as a whole class with the teacher having sets of cards in three piles: cards worth 1 point, 2 points and 3 points.

Individual pupils ask the teacher for a 1, 2 or 3 point card. When they think they have solved the equation, they ask for another card.

Continue until all the cards have been taken or some specified time limit has been reached.

One pile whole-class version

This is played as the Three pile whole-class version but pupils take cards at random from a mixed pile.

Individual version

The cards do not need to be cut out for this version. Each pupil solves as many equations as they can from the set of 36 in a specified time. Solutions are checked and points awarded as before.

The game can be played with the number of points awarded for a correct solution being the value of the solution.

A set of solutions is given below:

Card 1	$n = 3.5$	Card 13	$n = 5.2$	Card 25	$n = 4$
Card 2	$n = 1.5$	Card 14	$n = 4$	Card 26	$n = 0.6$
Card 3	$n = 2$	Card 15	$n = 29.5$	Card 27	$n = 19.3$
Card 4	$n = 9$	Card 16	$n = 0.6$	Card 28	$n = 5$
Card 5	$n = 18$	Card 17	$n = 5.8$	Card 29	$n = 0.7$
Card 6	$n = 0.6$	Card 18	$n = 2$	Card 30	$n = 1.7$
Card 7	$n = 2$	Card 19	$n = 6$	Card 31	$n = 4.48$
Card 8	$n = 4$	Card 20	$n = 8$	Card 32	$n = 35$
Card 9	$n = 5$	Card 21	$n = 2.7$	Card 33	$n = 3$
Card 10	$n = 3.3$	Card 22	$n = 6$	Card 34	$n = 143$
Card 11	$n = 108$	Card 23	$n = 0.9$	Card 35	$n = 21$
Card 12	$n = 241$	Card 24	$n = 1.5$	Card 36	$n = 34$

Ⓐ Number puzzles (p 260)

A1

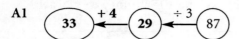

The number thought of was 33.

A2 (a) Puzzle 1 C; Puzzle 2 B

(b) Puzzle 1

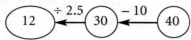

The number thought of was 12.

Puzzle 2

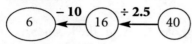

The number thought of was 6.

A3 (a) I think of a number.
- I divide by 4.
- I subtract 1.

The result is 1.5.

What was my number?

(b)

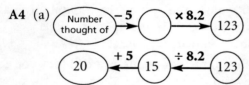

The number thought of was 10.

A4 (a)

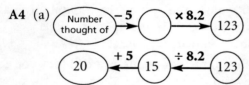

The number thought of was 20.

(b)

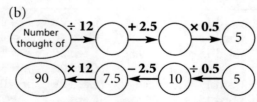

The number thought of was 90.

(c)

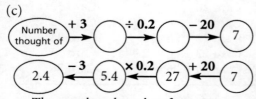

The number thought of was 2.4.

A5 Using positive whole numbers and zero gives the following possibilities.
- I add 10, I multiply by 1.
- I add 4, I multiply by 2.
- I add 2, I multiply by 3.
- I add 1, I multiply by 4.
- I add 0, I multiply by 6.

Extending to negative numbers and decimals gives an infinite number of possibilities.

A6 (a) The pupil's numbers and solutions

(b) The pupil's numbers and solutions

(c) The number thought of is the same as the result each time.

It can be explained by writing 'n' for the 'number thought of' and writing a simplified expression in terms of n as the result of each operation.

Number thought of n

I subtract 1.	$n - 1$
I multiply by 6.	$6(n - 1) = 6n - 6$
I add 3.	$6n - 3$
I divide by 3.	$2n - 1$
I add 1.	$2n$
I divide by 2.	n

Ⓑ Using letters (p 262)

B1 (a) A Y, B W, C V, D X

(b)

$$n \xrightarrow{\times 5} \bigcirc \xrightarrow{+7} 16$$

B2 (a) W D, X B, Y E, Z C

(b) I think of a number.
- I subtract 5.
- I multiply by 4.

The result is 8.

What was my number?

B3 (a) $3n + 4 = 108$ (b) $4.5n = 162$

(c) $\dfrac{n-2}{5} = 2.2$

B4 (a) $\dfrac{n}{3} + 6 = 21$ (b) $2.6(k - 5) = 65$

ℂ Solving equations (p 263)

C1 (a) $n = 12$ (b) $h = 10$ (c) $p = 21$
(d) $q = 71$ (e) $t = 16.8$ (f) $s = 19$

C2 (a) $n = 7.4$ (b) $p = 3.9$ (c) $q = 32$
(d) $x = 3.6$ (e) $y = 22.1$ (f) $z = 9.6$

C3 $(6 - 3) \div 5 = 3 \div 5 = 0.6$; the pupil's equations with solution $n = 6$

C4 The pupil's equations with solution $y = 1.5$

C5 (a) $m = 23$ (b) $p = 0.7$
(c) $s = 9.5$ (d) $t = 0.05$

What progress have you made? (p 264)

1 (a)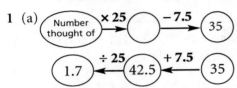

The number thought of was 1.7.

(b)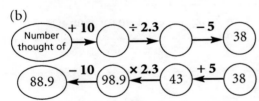

The number thought of was 88.9.

2 A Z, B W, C Y

3 (a) $z = 20$ (b) $p = 18$ (c) $y = 96$
(d) $q = 3.7$ (e) $x = 5.31$

Practice booklet

Section A (p 85)

1 (a) 32 (b) 9 (c) 2.8 (d) 248.4

2 (a) The pupil's result and solution
(b) The pupil's solution

(c) The result is always twice the number thought of. It can be explained by writing 'n' for the 'number thought of' and writing a simplified expression in terms of n as the result of each operation.

Number thought of n
I add 1 $n + 1$
I multiply by 4 $4(n + 1) = 4n + 4$
I add 6 $4n + 10$
I divide by 2 $2n + 5$
I subtract 5 $2n$

Section B (p 85)

1 $54n = 378$

2 (a) $7n + 3 = 206$ (b) $\frac{n-13}{4} = 6.1$

3 (a) $7(y - 3) = 21$ (b) $\frac{a}{5} + 3 = 18$
(c) $\frac{3.2p}{7} = 4$ (d) $5.1(m + 9) = 102$
(e) $\frac{x-10}{3} + 1 = 3$

Section C (p 86)

1 (a) $w = 11$ (b) $n = 8$ (c) $p = 31$
(d) $x = 15.3$ (e) $a = 77$ (f) $y = 39$
(g) $p = 23$ (h) $h = 16.2$ (i) $t = 152$

2 (a) $x = 5.1$ (b) $a = 4.3$ (c) $y = 50$
(d) $m = 4.2$ (e) $w = 15.3$ (f) $h = 11.5$
(g) $k = 13.5$ (h) $m = 216.9$
(i) $n = 46.4$

3 The pupil's equations with solution $x = 4$

4 The pupil's equations with solution $p = 2.5$

5 The pupil's equations with solution $n = 7.1$

6 (a) $x = 16$ (b) $d = 14$ (c) $y = 17.9$
(d) $m = 7$ (e) $b = 16.7$ (f) $n = 16$

*7 (a) $s = 12.9$ (b) $r = 1.5$
(c) $q = 2.4$ (d) $p = 8.5$

*8 $x = 8$

Quadrilaterals

The names and properties of the square, rectangle, parallelogram, rhombus, trapezium, kite and arrowhead are revised or introduced. Pupils see how special types of quadrilateral can be made up from certain types of triangle. The sum of the interior angles of a quadrilateral is established and used to find missing angles and as part of an exercise on accurate drawing. The fact that some types of quadrilateral are special cases of others is briefly explored.

p 265	**A** Special quadrilaterals	
p 268	**B** Quadrilaterals from triangles	
p 269	**C** Angles of a quadrilateral	
p 272	**D** Accurate drawing	
p 272	**E** Quadrilaterals from diagonals	Drawing a quadrilateral given information about its diagonals
p 273	**F** Stand up if your drawing …	Quadrilaterals which are special cases of other types
p 273	**G** Always, sometimes, never …	Identifying types of quadrilateral

---

Essential

Square dotty paper, sheet 164
Scissors, angle measurers, compasses

Practice booklet pages 88 and 89

Special quadrilaterals (p 265)

> Square dotty paper

◊ Page 265 provides an opportunity to lead a discussion to find out how many of the quadrilaterals' names and properties pupils know and to fill in any gaps in their knowledge.

One teacher reported 'I organised pupils into groups of four or five and gave them 30 minutes to prepare a presentation for the class on their particular shape. Each contained an accurate drawing of the shape and responses to the seven questions on the page. This worked well!'

A6–8 These questions bring in area and are appreciably harder than the rest.

B Quadrilaterals from triangles (p 268)

> Sheet 164, scissors

Although this section takes time it provides consolidation: pupils have to recognise triangle types and the special quadrilaterals they produce. It also offers practice in exploring all possibilities and can help pupils visualise quadrilaterals as built up from triangles. Most parts of B8 (and of section E) depend on this last idea.

◊ You may need to revise the different types of triangles before starting this section. Remember that some types of quadrilaterals are special cases of others (a point dealt with more fully later). So, for example, if a pupil creates a rhombus in B3 and labels it 'parallelogram' that is not wrong, but you could ask 'What special kind of parallelogram?'

C Angles of a quadrilateral (p 269)

> Angle measurers

◊ You could remind pupils of the proof that the angles of a triangle add up to 180°, in either of these versions

By splitting up a quadrilateral into two triangles, we *prove* that its angles add up to 360°. Make sure that pupils appreciate the difference between proving and merely verifying by measurement.

D Accurate drawing (p 272)

> Angle measurers, compasses

E Quadrilaterals from diagonals (p 272)

◊ Some pupils may need reassurance that the sketch is not trying to indicate the *shape* of any of the quadrilaterals: it merely shows how the vertices are lettered.

F Stand up if your drawing ... (p 273)

◊ It's probably best to start this section with the pupils' books shut.
We define a parallelogram as any quadrilateral that has two pairs of parallel sides. So a rectangle is a special kind of parallelogram.

Similarly,
a parallelogram is a special kind of a trapezium,
a rhombus is a special kind of a parallelogram,
a square is both a special rhombus and a special rectangle.

Hence,
a square is a special parallelogram,
a rhombus is a special trapezium,
and so on.

Introduce these ideas through discussion before going on to the following activity.

Six pupils sit on chairs facing the front of the class each holding one of these drawings.

(isosceles trapezium, trapezium with two right angles, and non-special cases of these: rectangle, parallelogram, rhombus, square).

Say to the group 'Stand up if your drawing is a rectangle.' If the square person doesn't stand up, ask the class if there is anyone not standing up who should be and, by inviting explanations, check that the special case idea has been understood.

Repeat the process with 'Stand up if your drawing ...
        ... is a rhombus'
        ... is a square'
and so on. You can also extend the idea to properties of their drawings:
'Stand up if your drawing ...
        ... has reflection symmetry'
        ... has at least one right angle'
and so on.

After you have done these activities, pupils can decide what the teacher asked for in the photographs shown on the page.

G Always, sometimes, never ... (p 273)

◊ Each numbered box describes a type of quadrilateral. Through teacher-led discussion, the class has to decide which. The discussion should not be hurried: pupils should have time to suggest quadrilaterals and comment on others' suggestions.

A Special quadrilaterals (p 265)

A1

A2 (a) Square

(b) Rhombus

(c) Rhombus

(d) Parallelogram

(e) Rhombus

(f) Parallelogram

A3

(a) (b)

A4

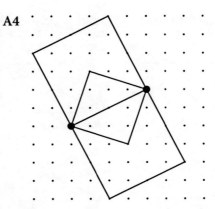

A5 The pupil's drawing of a rhombus (There are infinitely many possible.)

A6

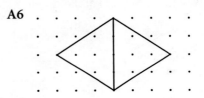

A7 Here are two possibilities. There are others.

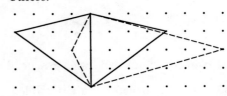

A8

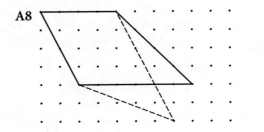

B Quadrilaterals from triangles (p 268)

B1 These quadrilaterals are possible.

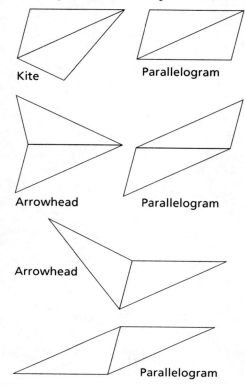

B2 You could not make an arrowhead.

B3 These quadrilaterals are possible.

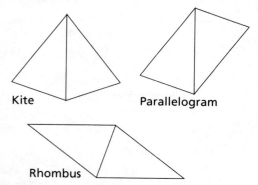

B4 A kite would not have been possible but an arrowhead would.

B5 Only this rhombus is possible.

B6 (a) These quadrilaterals are possible.

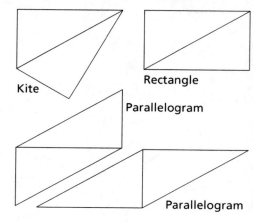

(b) These triangles are possible.

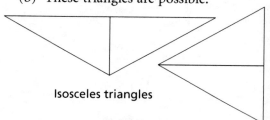

Isosceles triangles

B7 The rectangle and kite would have become a square. Only one isosceles triangle would have been possible, a right-angled one.

B8 (a) $\frac{1}{3}$ (b) $\frac{1}{4}$ (c) $\frac{1}{2}$
    (d) $\frac{3}{8}$ (e) $\frac{1}{7}$

C Angles of a quadrilateral (p 269)

C1 360°

C2 $a = 140°$ $b = 72°$ $c = 65°$

C3 $a = 222°$ $b = 100°$ $c = 112°$
    $d = 115°$ $e = 65°$ $f = 30°$
    $g = 92°$ $h = 140°$ $i = 105°$
    $j = 67°$ $k = 95°$ $l = 80°$
    $m = 55°$ $n = 100°$

C4 $a = 87°$ $b = 101°$ $c = 152°$ $d = 84°$

C5 $a = 135°$ $b = 45°$ $c = 60°$ $d = 45°$
    $e = 30°$ $f = 30°$ $g = 20°$ $h = 15°$
    $i = 30°$ $j = 24°$ $k = 18°$

D Accurate drawing (p 272)

D1 The pupil's drawings

D2 (a) 5.1 cm (b) 14.6 cm (c) 11.1 cm

D3 (a) 58° (b) 206°

D4 The pupil's drawing

D5 The pupil's drawing

D6 The pupil's drawing

E Quadrilaterals from diagonals (p 272)

E1 (a) A square (b) A kite
  (c) A rectangle (d) A rhombus
  (e) A parallelogram
  (f) An ('isosceles') trapezium
  (g) A (long thin) rectangle
  (h) A parallelogram

F Stand up if your drawing ... (p 273)

The teacher said:

1 'Stand up if your drawing is a parallelogram.'
2 'Stand up if your drawing has just two lines of reflection symmetry.'
3 'Stand up if your drawing is a trapezium' (or, possibly, '... if your shape is a quadrilateral').

G Always, sometimes, never ... (p 273)

1 A rhombus
2 A rectangle
3 A square
4 A kite
5 A rhombus
6 A trapezium
7 A kite
8 A parallelogram

What progress have you made? (p 274)

1 The pupil's drawings

2 A kite, an arrowhead

3 A square

4 $a = 92°$ $b = 240°$ $c = 55°$

5 The pupil's drawings

6 Square, rhombus, rectangle

Practice booklet

Section B (p 88)

1 The pupil's drawings, for example

(a)   (b)

(c) 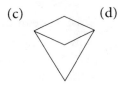   (d)

(e)

Section C (p 88)

1 $a = 98°$ $b = 117°$ $c = 102°$
  $d = 48°$ $e = 95°$ $f = 68°$
  $g = 244°$ $h = 16°$ $i = 144°$
  $j = 81°$

2 $a = 36°$ $b = 20°$ $c = 28°$
  $d = 18°$ $e = 25°$ $f = 20°$

Section D (p 89)

1 The pupil's accurate drawings

Negative numbers 2

This unit revises addition and subtraction of negative numbers and introduces multiplication and division.

Practice booklet page 90

Ⓐ Addition and subtraction patterns (p 275)

This approach to addition and subtraction complements the approach used in unit 18 'Negative numbers 1'. The idea of extending number patterns is used for multiplication in the next section.

◊ You could start by asking pupils to imagine that they do not know any rules for adding or subtracting negative numbers (which may well be the case anyway!).

Complete the next two lines of pattern A and notice that the result goes down by 1 each time. Use this fact to extend the pattern and add two or three more lines beyond those given.

Now ask pupils, working in pairs, to do the same for each of the other patterns. Help them to appreciate that the rules for adding and subtracting negative numbers are chosen so that the results 'fit in' with those for positive numbers.

Ⓑ Multiplication (p 276)

◊ It is probably best to reproduce the incomplete table on the board or an OHP transparency.

You could start by asking pupils how the table should be extended to the left. In row '4', for example, the numbers go down by 4 each time as you go to the left, so the row continues $^-4$, $^-8$, ...

Then work on extending the columns downwards, starting with columns '4', '3', ...

◊ Point out the square numbers on the diagonal starting with 16. Each square number (apart from 0) appears twice in the table and thus has two square roots, one positive and one negative.

◊ You could also ask high attainers to think about the consequences of choosing a different rule for multiplying two negative numbers. Most pupils will see that $3 \times {}^-2$ can be thought of as $^-2 + {}^-2 + {}^-2 = {}^-6$. However, if $^-3 \times {}^-2$ were also $^-6$, then $^-6 \div {}^-2$ would have two answers, 3 or $^-3$.

C **Division** (p 277)

D **Using inverse operations** (p 278)

A **Addition and subtraction patterns** (p 275)

A1 (a) 5 (b) 9 (c) 16
(d) $^-4$ (e) 16

A2 (a) 4 (b) $^-5$ (c) 5
(d) $^-5$ (e) 3

A3 (a) $^-4$ (b) 1 (c) $^-16$
(d) 13 (e) 3

A4 Each of the additions below can be written as another addition and in two ways as a subtraction.
$$^-2 + 3 = 1 \qquad ^-4 + 7 = 3$$
$$^-5 + 1 = {}^-4 \qquad ^-5 + 3 = {}^-2$$
There are 16 possibilities altogether.

B **Multiplication** (p 276)

B1 (a) $^-21$ (b) 21 (c) 30
(d) $^-24$ (e) 24

B2 (a) $^-30$ (b) 24 (c) $^-30$ (d) $^-48$

B3 $^-3$

B4 32

B5 (a) $^-15$ (b) $^-12$ (c) 23
(d) $^-36$ (e) 19 (f) 34

B6 Clare was not right. As well as 8, Kirsty's number could have been $^-8$.

B7 4 or $^-4$

*****B8** 6 or $^-8$

C **Division** (p 277)

C1 $^-20 \div 4 = {}^-5$, $^-20 \div {}^-5 = 4$

C2 $18 \div {}^-6 = {}^-3$, $18 \div {}^-3 = {}^-6$

C3 negative/positive = negative
positive/negative = negative
negative/negative = positive

C4 (a) $^-4$ (b) 10 (c) $^-4$ (d) $^-5$
(e) 8 (f) 4 (g) $^-7$ (h) $^-5$

C5 14

C6 (a) $^-9$ (b) $^-11$ (c) 16 (d) 21
(e) 7 (f) 2

C7 (a) $^-2$

(b) (i) $+ 6$, $\div {}^-3$, $+ 3$, $\times {}^-2$
(ii) $\div {}^-3$, $\times {}^-2$, $+ 3$, $+ 6$
(or $+ 6$, $+ 3$)
(iii) $\times {}^-2$, $+ 6$, $+ 3$, $\div {}^-3$
(or $\times {}^-2$, $+ 3$, $+ 6$, $\div {}^-3$)
(iv) $+ 3$, $\times {}^-2$, $+ 6$, $\div {}^-3$
(or $\div {}^-3$, $+ 3$, $\times {}^-2$, $+ 6$)

C8 (a) $\dfrac{24}{3} - \dfrac{^-12}{^-4}$ (b) $\dfrac{^-12}{3} - \dfrac{24}{^-4}$

(c) $\dfrac{^-12}{^-4} - \dfrac{24}{3}$ (d) $\dfrac{24}{^-12} - \dfrac{3}{^-4}$

D **Using inverse operations** (p 278)

D1 (a) 25
(b) (i) $^-7$ (ii) 2 (iii) 6

D2 (a) $^-4$
(b) (i) $^-31$ (ii) 17 (iii) 85

D3 (a) $^-7$ (b) 2 (c) $^-8$

D4 (a) $^-4$ (b) 25 (c) $^-14$

What progress have you made? (p 278)

1 (a) 12   (b) $^-9$ (c) $^-6$

2 (a) 28   (b) $^-27$ (c) $^-16$
   (d) $^-3$ (e) $^-8$ (f) $^-5$

3 (a) $^-2$ (b) 5

Practice booklet

Sections A and B (p 90)

1 (a) $^-13$ (b) $^-6$
   (c) 9   (d) 5

2 (a) $^-5$ (b) 15   (c) 13
   (d) $^-15$ (e) 9

3 (a) $^-24$ (b) $^-25$ (c) 27
   (d) $^-19$ (e) 44

4 3 or $^-3$

5 5 or $^-5$

Sections C and D (p 90)

1 (a) $^-1$ (b) 7   (c) $^-4$
   (d) 14   (e) $^-5$

2 (a) $^-6$ (b) 2   (c) 17
   (d) 14   (e) $^-1$

3 (a) $^-1$ (b) $^-5$ (c) 3

4 (a) 6   (b) 1

Comparisons 2

The mean is introduced using the idea of fairness.
There is some practical work in the unit, and a section that brings
together mean, mode and median.

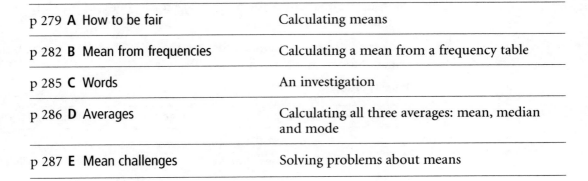

T	p 279 **A** How to be fair	Calculating means
T	p 282 **B** Mean from frequencies	Calculating a mean from a frequency table
	p 285 **C** Words	An investigation
T	p 286 **D** Averages	Calculating all three averages: mean, median and mode
	p 287 **E** Mean challenges	Solving problems about means

Essential	**Optional**
Packs of playing cards (ace to 10 only)	Newspapers Foreign language texts, other texts
Practice booklet pages 91 to 94	

A How to be fair (p 279)

T

◊ You could start by asking pupils in pairs to decide whether Sharon's or
Joshua's group did better, and why.

The mean for Sharon's group is 8 kg and for Joshua's it is 7 kg.

A6 This can generate discussion! Using the median reverses the order, and
pupils may bring in the idea of consistency.

Emphasise that the mean is not necessarily a possible number of points.

Mean tricks (p 281)

Packs of playing cards (ace to 10 only) or equivalent

◊ This game not only consolidates mental skills in calculating the mean but
can bring out other ideas (for example that the deviations from the mean
add up to zero). Since any number of cards (up to 7) can be used when
working out the mean, each hand potentially contains many calculations
of a mean.

Eavesdrop groups of pupils to hear their methods of finding the mean.
There could be class discussion of these methods. Schools have found
that the benefits pupils get from this game improve with playing as they
develop strategies.

B Mean from frequencies (p 282)

◊ It is worth spending some time on the meaning of the word 'frequency' in each situation. It is a common misconception to find the mean of the frequencies.

◊ Encourage pupils to set out their work logically. Although there are subroutines on a calculator, at this stage it is better to show all the steps so that errors can be traced.

B12 A common error is to assume that the mean for the whole population is equal to the mean of the means for the two parts.

Investigation

Newspapers (*Mirror* and *Guardian* suggested), foreign language texts

◊ Samples need to be of a reasonable size, say 20 to 30 sentences in each paper. In some schools this worked well as a homework task. Pupils were allowed to choose from a wider range of newspapers however.

C Words (p 285)

Optional: other texts

◊ This page is deliberately obscured so that pupils cannot count every word!

Some schools have used this successfully on books of their own choice. The activity takes about 15–20 minutes. If the piece is word-processed the teacher could use a word count.

D Averages (p 286)

◊ This section brings together the three measures of average met so far.
The word 'average' is used loosely to mean 'about the middle'.
For 'Mary is of average height' you could ask:

• Is 'average height' the average for the school year? the school? the population of the UK?

• What average could it be?

The 'average maximum daily temperature' would usually be understood as the mean of the maximum daily temperatures (probably rounded to the nearest degree).

In the sentence about the 'average family', the word 'average' has been attached to 'family' but really refers to an average amount of water.

E Mean challenges (p 287)

E1 It is worth discussing different ways of calculating an answer. For example, you could find the total weight before and after replacement or you could divide the increase in weight by the number in the team.

Ⓐ How to be fair (p 279)

A1 $27 \div 9 = 3$

A2 $120 \div 10 = 12$ goldfish

A3 (a) $40 \div 5 = 8$ pupils

(b) $36 \div 4 = 9$ pupils

A4 7 peppers per plant

A5 Ruth is right. The 24 peppers have to be shared between the four plants.

A6 Pat $32 \div 4$ or 8 points per game
Jon $39 \div 5$ or 7.8 points per game
Wayne $51 \div 6$ or 8.5 points per game

So you might consider Wayne to be the best points scorer.

A7 (a) 8L $40 \div 20 = 2$ cans
8N $55 \div 25 = 2.2$ cans
8N gets the prize.

(b) 8L $50 \div 20 = 2.5$ bottles
8N $60 \div 25 = 2.4$ bottles
8L gets the prize.
The pupil's explanation

(c) The pupil's own view and explanation
(8L average items = 4.5)
(8N average items = 4.6)

A8 Probably the best way to answer these questions is to calculate the mean load pulled by a member of the team.

Team A
mean load pulled = $2475 \div 9 = 275$ kg

Team B
mean load pulled = $1656 \div 6 = 276$ kg

Team C
mean load pulled = $1911 \div 7 = 273$ kg

(a) Team B   (b) Team C
The pupil's reasons

Ⓑ Mean from frequencies (p 282)

B1 (a) 5   (b) 15
(c) 1156 kg   (d) 77.1 kg

B2 (a) 34   (b) 124   (c) 3.6

B3 $(3 \times 4) + (4 \times 8) + (5 \times 5) + (6 \times 4) + (7 \times 3) + (8 \times 7) + (9 \times 5) = 215$
$215 \div 36 = 6.0$ (to 1 d.p.)

B4 (a) $(0 \times 5) + (1 \times 12) + (2 \times 13) + (3 \times 8) + (4 \times 2) = 70$

(b) $70 \div 40 = 1.75$

B5 $233 \div 40 = 5.825$

B6 $(8 \times 2) + (7 \times 3) + (6 \times 9) + (5 \times 1) = 96$
$96 \div 15 = 6.4$

B7 $222 \div 104 = 2.1$ (to 1 d.p.)

B8 1180p $\div 20 = 59$p

B9 $1435 \div 50 = 28.7$

B10 $8938 \div 177 = 50.5$ (to 1 d.p.)
The statement seems fair. Although you might only get 48 matches, it averages out at more than the 50 stated.

B11 (a) The mean of the pupil's set of five numbers

(b) The mean increases by 2.

(c) The mean increases by the amount you add to each number.

(d) $9760 + [(1 + 3 + 0 + 2 + 4) \div 5]$
$= 9762$

B12 (a) $265 \div 20 = 13.25$

(b) $420 \div 30 = 14$

(c) $685 \div 50 = 13.7$

C Words (p 285)

C1 One way is to find the mean number of words in a line, then multiply this by the number of lines.

Number of words in first six lines
18, 17, 17, 17, 18, 16

Mean number of words in first six lines
$= 103 \div 6 = 17. \dots$

Maximum number of words (assuming all lines of full length)
$= 17.16\dots \times 32 = 549.3$ (to 1 d.p.)

Minimum number of words (assuming some lines are very short)
$= 17.16\dots \times 28 = 480.7$ (to 1 d.p.)

So anything between 550 and 481 is reasonable, e.g. 515.

C2 $5000 \div 515 = 9.7$ (approx.)
i.e. about 9 or 10 pages

D Averages (p 286)

D1 (a) Median (b) Mode

D2 (a) Median 23, mean 26.5 (to 1 d.p.), mode 18

(b) The frequencies are too low for the modal age to be relevant.

(c)

Group	10–19	20–29	30–39	40–49	50–59
Frequency	11	9	6	3	1

The modal age group is 10–19.

D3 *Blackmouth*
Median 5
Mean $189 \div 31 = 6.1$ (to 1 d.p.)
Mode 4 and 5 equal
Bournepool
Median 7
Mean $178 \div 31 = 5.7$ (to 1 d.p.)
Mode 7

Although the mean number of hours of sunshine is less for Bournepool (5.7 compared with 6.1), the median is higher. The mode of 7 hours at Bournepool is more than that of the two most frequent amounts of sunshine at Blackmouth (4 and 5 hours).

E Mean challenges (p 287)

E1 Total weight before $= 7 \times 58 = 406$
after $= 406 - 40 + 54 = 420$
new mean $= 420 \div 7 = 60\,\text{kg}$
or
Increase in weight $= 14\,\text{kg}$
so increase in mean $= 14 \div 7 = 2\,\text{kg}$
new mean $= 58 + 2 = 60\,\text{kg}$

E2 Increase $= 7 \times 2 = 14\,\text{kg}$
so new player's weight $= 43 + 14 = 57\,\text{kg}$

E3 $(4 \times 143 + 140) \div 5 = 676 \div 5 = 135.2\,\text{cm}$

E4 $(6 \times 45) + (5 \times 51) = 525$
$525 \div 11 = 47.7\,\text{kg}$ (to 1 d.p.)
Note that it is not the mean of the two means.

E5 $(5 \times 46) - (40 \times 4) = 70$

What progress have you made? (p 287)

1 $94 \div 7 = 13.4°C$ (to 1 d.p.)

2 $70 \div 25 = 2.8$ eggs

Practice booklet

Section A (p 91)

1 (a) Kathy did better.
Her mean score was 3.5,
Mark's mean score was 3.4.

(b) Mark would have done better.
His mean would then be $16 \div 4 = 4$ skittles.

2 (a) The girls' mean pocket money is
£40.00 ÷ 10 = £4.00.

   (b) The boys' mean pocket money is
£31.60 ÷ 8 = £3.95.

3 52 ÷ 13 = 4 letters per word

4 48.5p

5 (a) The pupil's answer, for example
5, 7, 9 or 3, 5, 13

   (b) The total of the three numbers must
be 21 which is an odd number.
When even numbers are added the
answer is always even.

6 The pupil's answer, but the total of the
five numbers must be 47

Section B (p 92)

1 (3 × 28) + (1 × 26) = 110
110 ÷ 4 = 27.5 teeth

2 (1 × 50) + (2 × 25) + (3 × 20) + (4 × 3)
+ (5 × 2) = 182
182 ÷ 100 = 1.82 people per car

3 (4 × 5) + (5 × 1) + ... + (14 × 4) = 440
460 ÷ 50 = 9.2 eggs per lay

4 (a) 5370 ÷ 25 = 214.8 grams

   (b)

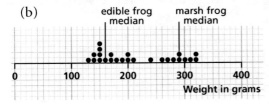

   (c) 10

   (d) 2880 ÷ 10 = 288 grams

   (e) 2490 ÷ 15 = 166 grams

Section E (p 93)

1 (10 × 47.2) + (20 × 42.4) = 1320
1320 ÷ 30 = 44 kg

2 (a) 15.43 ÷ 8 = 1.93 m

   (b) The pupil's estimate

   (c) 16.89 ÷ 9 = 1.88 m

3 (a) 66 ÷ 10 = 6.6 goals

   (b) 61 ÷ 10 = 6.1 goals

   (c) No
The answers to parts (a) and (b)
would suggest otherwise.

   (d) Mean goal margin = 3.7 goals
The answers to parts (a) and (b) do
not predict this.

4 The one who joined was 18 years
younger than the one who left.
(42 × 6 − 39 × 6 = 18)

*5 There are eight sets:
1, 2, 3, 5, 6, 11, 14
1, 2, 3, 5, 7, 10, 14
1, 2, 3, 5, 8, 9, 14
1, 2, 4, 5, 6, 10, 14
1, 2, 4, 5, 7, 9, 14
1, 3, 4, 5, 6, 9, 14
1, 3, 4, 5, 7, 8, 14
2, 3, 4, 5, 6, 7, 15

Review 4 (p 288)

1 (a) $5a = 20 = 5(a + 4)$

(b) $6b - 12 = \mathbf{6(b - 2)}$

(c) $6c - 20 = \mathbf{2(3c - 10)}$

(d) $15d + 24 = \mathbf{3(5d + 8)}$

2 (a) 400   (b) 0.036

(c) 8.1   (d) $30 \times 0.04 = 1.2$

3 $a = 66°$, $b = 66°$, $c = 73°$, $d = 107°$
(all with reasons)

4 Alex £580.50, Bharat £464.40,
Chris £245.10

5 (a) 13   (b) 1.4

6 Here is one possibility for each.

(a)   (b)

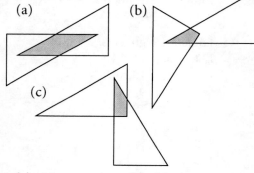

(c)

7 (a) 15   (b) 19

8 (a) Red 30%, blue 24%, green 11%,
white 22%, black 8%, other 5%

(b) The pupil's pie chart

9 (a) C, D   (b) A, F   (c) none

10 (a) $3x - 1$ (b) $15y^2$ (c) $p - 8$

11 3.1

12 (a) ⁻4   (b) 29   (c) ⁻1

(d) 9   (e) ⁻5

13 $a = 102°$, $b = 103°$, $c = 116°$, $d = 112°$
(all with reasons)

14 551 is not prime. $551 = 19 \times 29$

15 (a) 53°   (b) 75°   (c) 49°

16

Fraction	$\frac{4}{5}$	$\frac{7}{100}$	$\frac{12}{25}$	$\frac{13}{20}$	$\frac{17}{50}$	$\frac{1}{20}$	$\frac{1}{8}$
Decimal	**0.8**	0.07	**0.48**	0.65	**0.34**	**0.05**	0.125
Percentage	**80%**	**7%**	48%	**65%**	**34%**	5%	**12.5%**

17 420

18 (a) $125\,\text{cm}^2$ (b) $271\,\text{cm}^2$

19 The pupil's drawings
Angle B = 94°, angle C = 101°,
angle D = 57°, angle R = 65°,
PS = 6.5 cm, SR = 5.8 cm

Mixed questions 4 (Practice booklet p 95)

1 (a) $3p + 15$ (b) $4(q - 3)$

(c) $5r - 2$ (d) $\frac{1}{2}s + 7$

2 (a) 245   (b) 2.597   (c) 0.025 97

(d) 10.6   (e) 0.245   (f) 0.106

3 138°

4 (a) £3.25   (b) £9.60   (c) £10.50

(d) £8.40   (e) £17.10   (f) £5.00

5 (a) Trapezium, kite

(b) Rectangle, rhombus, square

6 (a) ⁻9   (b) 12

7 79.4 kg

8 42 m

9 (a) $x = 24$ (b) $x = 5.5$ (c) $x = 128$

(d) $x = 5$ (e) $x = ⁻3$ (f) $x = ⁻8$

10 15

Know your calculator

The priority rules for multiplication, division, addition and subtraction are introduced by investigating how a scientific calculator evaluates expressions. The unit also covers the use of brackets, memory, square and square root and the 'change sign' key.

Scientific calculators are essential.

Essential

Scientific calculators (one for each pupil, for all sections of the unit)

Practice booklet pages 97 to 100

Ⓐ Order of operations (p 291)

◊ You could begin by asking pupils to predict what they think their calculators will give for each set of key presses on page 291.

Pupils find the result of each set of key presses (remind them they need to press the '=' key or 'ENTER' key at the end) and try to describe the rules they think the calculator uses to evaluate expressions that include any combination of the four operations. They should try their rules out on their own expressions. Working in groups, each group could try to produce a clear statement of the rules they think the calculator follows.

A brief statement of the rules could be:

- You multiply or divide before you add or subtract.
- Otherwise, work from left to right.

◊ Some calculators use the symbols * and / for × and ÷ respectively.

A1–2 A calculator can be used to check results.

A3 Pupils could consider how many different results are possible. They could make up their own puzzles like this for someone else to solve.

Ⓑ Brackets (p 292)

Ⓒ A thin dividing line (p 292)

D Complex calculations (p 294)

The memory or an intermediate '=' can be used instead of brackets. The choice is a purely personal one.

E Squares (p 294)

In more complex calculations such as those at the end of E7, it is not necessary for pupils to use a continuous string of key presses (although some might like the challenge!). Some pupils will feel much more confident if they can write down intermediate steps.

F Square roots (p 296)

G Negative numbers (p 296)

A Order of operations (p 291)

A1 (a) 24   (b) 33   (c) 43   (d) 6
     (e) 5   (f) 9   (g) 16   (h) 25

A2 (a) 5   (b) 6   (c) 20   (d) 10
     (e) 10   (f) 4   (g) 32   (h) 12
     (i) 5

A3 (a) $24 \times 8 - 2$ (b) $24 + 8 \div 2$
     (c) $24 - 8 \times 2$ (d) $24 \div 8 \div 2$

B Brackets (p 292)

B1 (a) 35   (b) 47   (c) 5   (d) 11
     (e) 5   (f) 3   (g) 5   (h) 7

B2 (a) 11   (b) 3   (c) 6   (d) 3
     (e) 11   (f) 9   (g) 2   (h) $^-1$
     (i) 27

B3 (a) No   (b) Yes   (c) No   (d) Yes
     (e) No   (f) Yes   (g) No   (h) Yes
     (i) No   (j) No   (k) Yes   (l) No

C A thin dividing line (p 292)

C1 A and C
    B and G
    D and F
    H and I

C2 (a) $60 + \dfrac{3}{2}$ (b) $\dfrac{23 - 7}{4}$

     (c) $\dfrac{100 + 5}{3}$ (d) $4 - \dfrac{6}{5}$

     (e) $\dfrac{5}{3 - 1}$ (f) $\dfrac{18}{9} - 1$

     (g) $4 + \dfrac{12}{5 - 2}$

C3 (a) 7   (b) 12.5 or $12\frac{1}{2}$ (c) 1
     (d) 4   (e) 5   (f) 2
     (g) 7   (h) 10

C4 (a) 5   (b) 5   (c) 7
     (d) 35   (e) 5

C5 (a) 4   (b) 7

C6 (a) 1.5 (b) 60 (c) 5

C7 (a) (i) 19 (ii) 27
(iii) 9 (iv) 5
(b) The pupil's description

C8 (a) 21 (b) 5 (c) 2

C9 A and C

C10 [7] [−] [1] [=] [÷]
[(] [2] [+] [1] [)]

or

[(] [7] [−] [1] [)] [÷]
[(] [2] [+] [1] [)]

D Complex calculations (p 294)

D1 (a) 19.629 (b) 2.125 (c) 17
(d) 10.4

D2 (a) 11.25 (b) 9 (c) 2.5
(d) 22.1 (e) 3.8 (f) 0.5
(g) 3.9 (h) 3 (i) 31

E Squares (p 294)

E1 (a) 25 (b) 50 (c) 42
(d) 25 (e) 145 (f) 12
(g) 2 (h) 4

E2 (a) 100 (b) 400 (c) 2
(d) 9 (e) 8

E3 (a) [5] [×] [7] [x^2]

(b) [(] [5] [+] [7] [)] [x^2]

(c) [1] [0] [0] [÷] [5] [x^2]

E4 (a) 72 (b) 1 (c) 81 (d) 2
(e) 14 (f) 33 (g) 6.5

E5 B, D and E

E6 (continued)

[1] [5] [+] [9] [x^2] [=]
[÷] [4] [x^2]

or

[(] [1] [5] [+] [9] [x^2] [)]
[÷] [4] [x^2]

E7 (a) 5.76 (b) 5 (c) 1.5
(d) 36 (e) 90 (f) 7.44
(g) 3 (h) 10.5 (i) 10.388

E8 [5] [×] [6] [^{2}] [+] [4] [÷] [3]

F Square roots (p 296)

F1 (a) 5 (b) 19 (c) 48 (d) 2

F2 (a) 3.5 (b) 11.56
(c) 6.002 (d) 7
(e) 18.7 (f) 27.5625

F3 One possibility is given here for each part.
(a) $9 + \sqrt{4} \div 1$ (b) $9 + 4 \div \sqrt{1}$
(c) $1 + 9 \div \sqrt{4}$ (d) $\sqrt{9} + 1 \div 4$
(e) $1 + \sqrt{9} \div 4$ (f) $9 + 1 \div \sqrt{4}$

G Negative numbers (p 296)

G1 The pupil's key presses

G2 (a) −3.75 (b) −11.25
(c) 0.8375 (d) −8.1

G3 (a) 64.1 (b) 59.6

G4

¹1	0	²5	■	³1
2	■	⁴3	2	8
■	■	3	■	■
⁵4	7	6	■	⁶7
9	■	⁷1	9	8

What progress have you made? (p 297)

1 (a) 39   (b) 38   (c) 2   (d) 1

2 (a) 33.4   (b) 18
   (c) 0.523 437 5   (d) 5

3 (a) 27.76   (b) 4.5

4 (a) 24.04   (b) ⁻5

Practice booklet

Section A (p 97)

1 (a) 12   (b) 50   (c) 27   (d) 2
   (e) 12   (f) 18   (g) 12   (h) 11
   (i) 29

2 (a) 4   (b) 3   (c) 6
   (d) 6   (e) 12   (f) 5
   (g) 7   (h) 33   (i) 3

3 (a) $36 \div 9 \times 3$ (b) $36 - 9 \div 3$
   (c) $36 - 9 \times 3$ (d) $36 \times 9 \div 3$

4 (a) $16 + 8 \div 4$ (b) $16 + 4 \times 8$
     or $8 \div 4 + 16$ or $4 \times 8 + 16$
   (c) $8 + 16 \div 4$ (d) $4 \times 8 \div 16$
     or $16 \div 4 + 8$ or $8 \div 16 \times 4$
     or $16 - 8 + 4$ or $4 - 16 \div 8$

Sections B and C (p 98)

1 (a) $3 \times 3 \times 4$ (b) $4 \times (5 + 4)$
   (c) $4 + 4 \times 8$ (d) $(20 - 2) \times 2$
   (e) $24 + 3 \times 4$ (f) $4 \times (12 - 3)$

2 A and F, B and H, C and G, D and E

3 Expressions A, B, D and E do not need brackets.

4 A and G, B and H, C and E, D and F

5 (a) 4   (b) 12   (c) 7.5   (d) 1
   (e) 4   (f) 5   (g) 3   (h) 4

Section D (p 99)

1 (a) 6.5   (b) 24.3   (c) 8.5

2 (a) 5   (b) 3.9   (c) 44.8
   (d) 29.6   (e) 32   (f) 6.65

Section E (p 99)

1 (a) 5   (b) 36   (c) 39   (d) 40
   (e) 94   (f) 8   (g) 4   (h) 9

2 (a)
   (b)
   (c)
   (d)

3 (a) 10.8   (b) 42.25   (c) 5
   (d) 1.75   (e) 4.375   (f) 8.56
   (g) 6   (h) 4.5   (i) 1.4
   (j) 25.5   (k) 13   (l) 8.8

*4 $\dfrac{a^2 - b^2}{a - b} = a + b$

Sections F and G (p 100)

1 (a) ⁻14.5   (b) 1.7   (c) 51.894
   (d) 5.7   (e) 28.7   (f) 15.265 625

2 (a) 5.98   (b) ⁻46.44   (c) 461
   (d) 703   (e) ⁻39.25   (f) ⁻7.7
   (g) 5   (h) 4.5   (i) 6
   (j) 78.06   (k) 8.4

*3

¹7	²1 . 7	³5		⁴2
	1		⁵3 . 3	6
⁶2	9 . ⁷1	9		2
5		⁸6	6 . ⁹1	5
¹⁰2 . 4	5		0	
8		¹¹8 . 3	8	4

Three dimensions

T

p 298 **A** Drawing three-dimensional objects

p 298 **B** Views

p 300 **C** Volume of a cuboid

T

p 302 **D** Prisms

p 302 **E** Volume of a prism

T

p 305 **F** Nets

p 307 **G** Surface area

Essential	**Optional**
Linking cubes	Sheet 170
Triangular dotty paper	Collection of three-dimensional shapes, e.g. cartons

Practice booklet pages 101 to 107

Ⓐ **Drawing three-dimensional objects** (p 298)

Linking cubes, triangular dotty paper

◊ If objects which are 'mirror images' of each other are counted as different, there are 8 objects which can be made with four cubes and 29 objects with five cubes.

Four cubes

 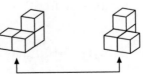

mirror images

Five cubes

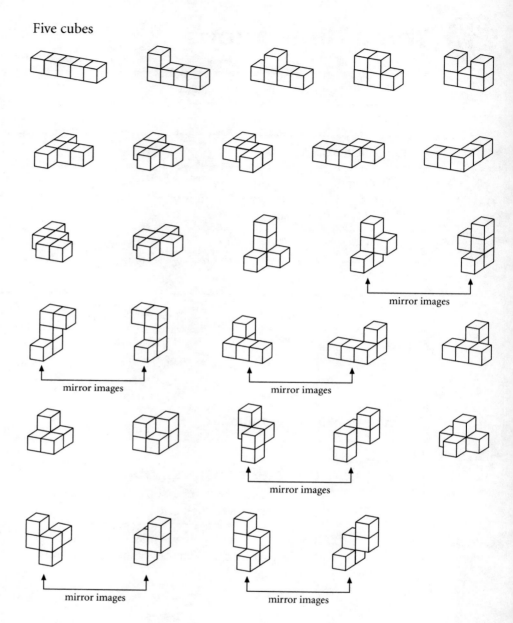

mirror images

mirror images

mirror images

mirror images

mirror images

mirror images

B **Views** (p 298)

The choice of 'front' and 'side' view is arbitrary.

C **Volume of a cuboid** (p 300)

Before doing work on volume it is a good idea to ask pupils how they would define 'cuboid'; for example, how would they describe a cuboid over the phone? This is a preparation for the more tricky question later of defining 'prism'.

D Prisms (p 302)

◊ Defining 'prism' is harder than defining 'cuboid'. This is a task which could be set to pupils working in pairs or groups. It is useful to have a number of three-dimensional shapes available to help clarify later what is and what is not a prism. Have some counter-examples ready; for example, the suggested definition 'same shape all through' can be met with a twisted pile of paper.

◊ The only prisms used at this stage are 'right' prisms, where a plane shape (the cross-section) is translated in a direction perpendicular to its plane.

E Volume of a prism (p 302)

F Nets (p 305)

Pupils should be encouraged to do as much as they can by visualising the three-dimensional shapes – they should cut out and fold nets only when they cannot visualise the shapes, or to check their thinking.

◊ Nets P, Q and R are on the optional resource sheet 170. Having discussed their sketches, perhaps first in small groups, and then as a whole class, pupils could make the solids.

◊ P square-based
pyramid

Q cube with
corner
sliced off

R 'house' or
cube with
triangular
prism

◊ Useful discussion questions are 'What is the same about each solid? What is different?' It is hoped that pupils will consider and discuss faces, vertices and edges, but do not force this.

G Surface area (p 307)

B Views (p 298)

B1 (a) (b)

B2 (a) Front Side

Top

(b) Front Side

Top

(c) Front Side

Top

B3 (a) A P, B R, C S, D Q
(b) View from T

B4 Mug A, G

Spoon B, D

Toilet roll C, E

Book F, H

or

C Volume of a cuboid (p 300)

C1 (a) $40\,cm^3$ (b) $60\,cm^3$ (c) $24\,cm^3$

C2 (a) $30\,cm^3$ (b) $48\,cm^3$
(c) $28\,cm^3$ (d) $36\,cm^3$

C3 $10.5\,cm^3$. You can see nine whole cubes and three halves.

C4 $12.25\,cm^3$

C5 (a) $36\,cm^3$ (b) $45\,cm^3$
(c) $16.75\,cm^3$ (d) $20.25\,cm^3$

C6 $a = 4\,cm$ $b = 9.6\,cm$ $c = 1.5\,cm$
$d = 4\,cm$ $e = 3.63\,cm$ (to 2 d.p.)

C7 For example, 10 by 20 by 24 cm,
10 by 16 by 30 cm, 12 by 16 by 25 cm;
volume = $4800\,cm^3$ in each case

E Volume of a prism (p 302)

E1 (a) $60\,cm^3$ (b) $98\,cm^3$ (c) $125\,cm^3$
(d) $40\,cm^3$ (e) $96\,cm^3$ (f) $123\,cm^3$

E2 (a) $48\,cm^3$ (b) $84\,cm^3$ (c) $186\,cm^3$

E3 (a) $60\,cm^3$ (b) $94.5\,cm^3$ (c) $180\,cm^3$
(d) $60\,cm^3$ (e) $108\,cm^3$

E4 $2.79\,m^3$

E5 $0.55\,m = 55\,cm$

E6 $40\,cm$

E7 $4000\,cm^2$

E8 $40\,m^2$

E9 The pupil's sketches

E10 The pupil's sketches

E11 The volume is multiplied by
(a) 2 (b) 4 (c) 8

*E12 7 cm by 14 cm by 21 cm
(width, height, length)

*E13 11 cm

F1 (a) The missing face could go on the net in any one of four positions.

(b)

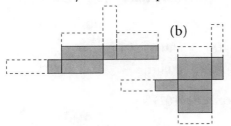

F2 (a) P and Q are nets of a cube, R is not.

(b) The arrangements which make a cube are

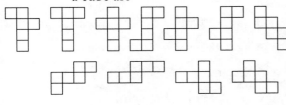

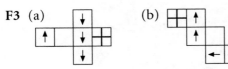

F3 (a)   (b)

F4 (a)

(b)

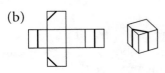

F5 (a) Square based pyramid

(b) Triangular prism

(c) Tetrahedron

(d) Triangular prism ('wedge')

(e) Hexagonal prism

F6 (a) There are several possibilities. Here are two.

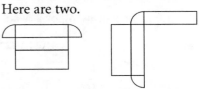

(b) There are several possibilities. Here is one.

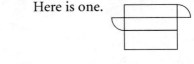

F7

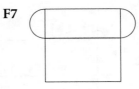

F8

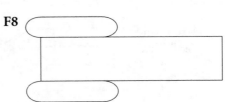

G **Surface area** (p 307)

G1 126 cm^2

G2 (a) 122 cm^2 (b) 202 cm^2
(c) 134.5 cm^2 (d) 35.5 cm^2

G3 Not necessarily. Pupils should give at least one example of a pair of cuboids with the same volume but different surface areas – e.g. 2 by 4 by 2 (surface area 40) and 8 by 2 by 1 (surface area 52).

G4 192 cm^2
(The missing edge length is 12 cm.)

G5 248.5 cm^2
(The missing edge length is 6.5 cm.)

G6 9 cm

G7 16 cm^3

What progress have you made? (p 308)

1 The pupil's drawing of a 5-cube object

2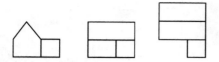

3 30 cm³

4 70 cm³

5 0.42 m (to 2 d.p.)

6 (a) Here is one possibility.

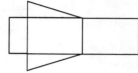

(b) Here is one possibility.

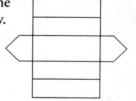

7 67 cm²

Practice booklet

Section B (p 101)

1 (a)   (b)

2 front side   top

3 A   B   C

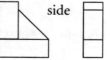

Section C (p 101)

1 (a) 208 cm³   (b) 675 cm³
(c) 1700 cm³   (c) 189 cm³

2 $a = 2$ cm, $b = 1.25$ cm, $c = 5$ cm

3 (a) 1 000 000 cm³
(b) 1000 litres   (c) 72 000 litres

Section E (p 102)

1 (a) 46 cm³   (b) 36 cm³
(c) 90 cm³   (d) 90 cm³

2 (a) 30 cm³   (b) 42 cm³

3 (a) 120 cm³   (b) 25 cm³

4 (a) 280 cm³   (b) 560 cm³

5 24 cm

6 $a = 2.5$ cm, $b = 9$ cm, $c = 3$ cm

Section F (p 104)

1 (a)   (b)

(c)   (d)

2 There are two different possible dice, one with the numbers 2 3 5 4 going clockwise around the number 1,

and the other with the numbers 2 3 5 4 going anticlockwise around the number 1.

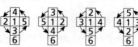

3 B, E, F

4 (a)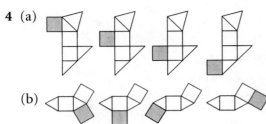

(b)

5 (a)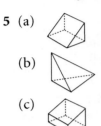

(b)

(c)

6 (a) Equilateral triangle

(b) Square

(c) Rhombus

(d) Regular hexagon

Section G (p 106)

1 (a) 14 cm   (b) 84 cm^2 (c) 104 cm^2

2 (a) $2 \times (4 \times 6) + (7 \times 20) = 188$ cm^2

(b) $2 \times (7 \times 2) + (6 \times 18) = 136$ cm^2

(c) $2 \times (3 \times 7) + (3 \times 20) = 102$ cm^2

3 (a) 28 cm   (b) 7 cm

(c) 196 cm^2 (d) 276 cm^2

4 (a) $(2 \times 96) + (5 \times 56) = 472$ cm^2

(b) $(2 \times 84) + (8 \times 50) = 568$ cm^2

(c) $(2 \times 84) + (6 \times 56) = 504$ cm^2

5 (a) (i) 6 cm^2 (ii) 24 cm^2 (iii) 54 cm^2

(b) (i) 1 cm^3 (ii) 8 cm^3 (iii) 27 cm^3

(c) 96 cm^2 and 64 cm^3

(d) 864 cm^2 and 1728 cm^3

6 (a) (i) 94 cm^2 (ii) 376 cm^2

(b) (i) 70 cm^2 (ii) 280 cm^2

The surface area of the second cube in each pair is 4 times the surface area of the first.

7 (a) (i) 70 cm^2 (ii) 630 cm^2

(b) (i) 40 cm^2 (ii) 360 cm^2

The surface area of the second cube in each pair is 9 times the surface area of the first.

Finding formulas

This is about finding a formula for the *n*th term of a sequence. At first the sequences are related to spatial patterns which can be analysed to give an expression for the *n*th pattern. The later part of the unit focuses on linear sequences of numbers.

T

p 309 **A** Maori patterns — Finding a formula relating to the *n*th pattern in a sequence of patterns

p 311 **B** Matchstick patterns — More sequences of patterns

T

p 312 **C** Sequences

p 313 **D** Sequences from rules — Substituting in a given formula for the *n*th term

T

T

p 313 **E** Finding a formula for a linear sequence

p 315 **F** Decreasing linear sequences

Optional
Squared paper
OHP, matches

Practice booklet pages 108 to 110

Maori patterns (p 309)

T

> Optional: squared paper

◊ The objective of the introduction is to bring out the different ways of tackling the task of finding a general rule. There is a sequence of patterns on page 309 which can be used as a basis for this.

You could set pupils (possibly working in pairs) the problem of finding how many crosses there will be in pattern 10, then 100 and to see if they can write a formula for the number in pattern *n*. It is important that everybody should have a go at this task before they hear anyone's answers.

◊ It is essential in the subsequent discussion to bring out the two ways of generalising: geometrical and numerical.

Geometrical – Pupils may have different ways of breaking up the pattern but this is probably the most common:

'Pattern 10 will have 10 crosses to the left, 10 to the right, 10 below and one in the middle, leading to $3n + 1$ for pattern n.

Numerical – 'Make a table of the number of crosses in order.

Pattern number	1	2	3	4	5
Number of crosses	4	7	10	13	16

The number of crosses goes up by 3 each time, so the formula must start with $3n$. Pattern 1 has 4 crosses, so the formula must be $3n + 1$.'

At this stage, pupils using the numerical approach are unlikely to see how to get from 'goes up in 3s' to '$3n$'. This is covered later in the unit.

In any case, the geometric approach is better in this section and the next because some of the patterns give quadratic sequences.

A3, A4 These questions give opportunities to discuss equivalent expressions. For example in question A3:

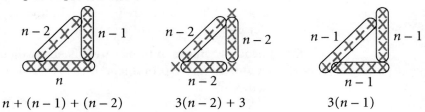

$n + (n - 1) + (n - 2)$ $3(n - 2) + 3$ $3(n - 1)$

B Matchstick patterns (p 311)

Optional: OHP, matches

C Sequences (p 312)

◊ You could start by asking pupils to find as many different ways as they can of continuing the two sequences given (1, 2, 4, … and 2, 4, 6, …).

Bring out the two different ways of giving a rule for a sequence. For example, the term-to-term rule for 2, 4, 6, … is 'add 2', and the position-to-term rule is 'multiply the position (term number) by 2'.

Of course, when using a term-to-term rule, you have to know the first term as well.

In case pupils are short on ideas for continuing the two sequences, here are some you can mention:

1, 2, 4, 8, 16, 32, … (multiply by 2)

1, 2, 4, 7, 11, 16, … (add 1, add 2, add 3, …)

1, 2, 4, 5, 7, 8, … (add 1, add 2, add 1, add 2, …)

1, 2, 4, 9, 23, 64, … (multiply by 3 and subtract 1,
multiply by 3 and subtract 2,
multiply by 3 and subtract 3, …)

2, 4, 6, 8, 10, … (add 2)

2, 4, 6, 10, 16, 26, …(add the previous two terms)

◊ Emphasise the fact that although there may be an 'obvious' rule for continuing a sequence, it does not follow that this is the only possible rule. However, pupils should assume that when asked for *the* rule for a sequence, the obvious one is intended.

◊ Explain that a linear sequence is so called because if the terms are plotted on a graph the points are in a straight line. (This point is returned to in a later unit.)

C4 Encourage pupils to think carefully about their methods and to try to avoid trial and improvement. It is a good idea for them to compare methods.

Ⅾ **Sequences from rules** (p 313)

ⅇ **Finding a formula for a linear sequence** (p 313)

◊ Give everybody a chance to see if they can find a rule for the nth term of each of the three sequences given. If pupils are struggling with B or C, a hint along these lines is useful:

Start with the sequence 1, 2, 3, 4, 5, …
for which the nth term is just n.

Choose a number to multiply every term by, say 4.

Notice that the new sequence, whose nth term is $4n$, goes up by 4 each time.

Choose a number to add to every term, say 3.

Notice that the new sequence still goes up in 4s, and this is true whatever is added or subtracted.

```
   1    2    3    4
  ↓(×4) ↓    ↓    ↓
   4    8   12   16
  ↓(+3) ↓    ↓    ↓
   7   11   15   19
```

◊ For the nth term of A, some pupils may produce an expression with '+3' in it. They are confusing the relationship across the table with the relationship downwards.

◊ Ask pupils to explain their own approaches to B and C, before focusing on the approach used at the top of page 314.

Decreasing linear sequences (p 315)

This work could be followed up by the activity 'Making a sequence' in 'Using a spreadsheet'.

Ⓐ **Maori patterns** (p 309)

A1 (a)

Pattern 4   Pattern 5

(b) Pattern 10 has four 'arms' each of 10 crosses and one cross in the middle.

(c)
Pattern number	1	2	3	4	5
Number of crosses	5	9	13	17	21

(d) (i) 41   (ii) 201   (iii) 401

(e) $4n + 1$

A2 (a)

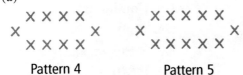

Pattern 4   Pattern 5

(b) Pattern 10 has two rows of 10 crosses and one cross at each end.

(c) 202

(d) $2n + 2$

(e) Pattern 27

A3 (a)

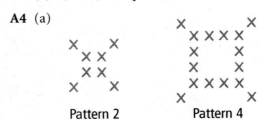

Pattern 3   Pattern 5

(b) 297

(c) $3(n - 1)$ or equivalent

A4 (a)

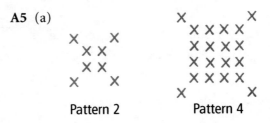

Pattern 2   Pattern 4

(b) Pattern 10 is an open square with 10 crosses on each side with an extra cross at each corner.

(c) 400

(d) $4n$

(e) Pattern 21

A5 (a)

Pattern 2   Pattern 4

(b) Pattern 10 is a 10 by 10 square of crosses with an extra cross at each corner.

(c) 104

(d) $n^2 + 4$

(e) Pattern 15

*A6 $n(n + 1)$ or equivalent

A7 The pupil's design, with formula

Ⓑ Matchstick patterns (p 311)

B1 $2n$

B2 $3n + 1$

B3 $5n + 1$

B4 $6n - 2$

B5 $8n + 7$

*B6 $2n(n + 1)$

Ⓒ Sequences (p 312)

C1 (a) Add 2, then 3, then 4, … and so on.
   (b) 28

C2 (a) (i) Add 5   (ii) 44
   (b) (i) Add 1, 3, 5, … (odd numbers)
      (ii) 51
   (c) (i) Add 2, 4, 8, … (double each time)
      (ii) 255
   (d) (i) Subtract $\frac{1}{4}$
      (ii) $1\frac{1}{4}$
   (e) (i) Multiply by 2
      (ii) 384
   (f) (i) Add the two previous terms
      (ii) 47

C3 (a) Linear   (b) Non-linear
   (c) Non-linear   (d) Linear
   (e) Linear

C4 Puzzle 1: 10th term is 49
Puzzle 2: 6th term is 17
Puzzle 3: 1st term is 7
Puzzle 4: 8th term is 30
Puzzle 5: 11th term is 49
Puzzle 6: 6th term is 15
Puzzle 7: 7th term is 41

Ⓓ Sequences from rules (p 313)

D1 (a) 1, 4, 7, 10, 13
   (b) The differences are 3, 3, 3, 3, …
      It is a linear sequence.

D2 (a) $4n + 2$ (b) $n + 4$ (c) $5n - 8$
   (d) $n^2 - 1$ (e) $8n - 3$

D3 (a) (i) 5, 10, 15, 20, 25, 30
      (ii) Linear
      (iii) 500
   (b) (i) 6, 11, 16, 21, 26, 31
      (ii) Linear
      (iii) 501
   (c) (i) 4, 8, 12, 16, 20, 24
      (ii) Linear
      (iii) 400
   (d) (i) 1, 5, 9, 13, 17, 21
      (ii) Linear
      (iii) 397
   (e) (i) ⁻6, ⁻3, 0, 3, 6, 9
      (ii) Linear
      (iii) 291
   (f) (i) 1, 4, 9, 16, 25, 36
      (ii) Non-linear
      (iii) 10 000
   (g) (i) 49, 48, 47, 46, 45, 44
      (ii) Linear
      (iii) ⁻50
   (h) (i) 2, 5, 10, 17, 26, 37
      (ii) Non-linear
      (iii) 10 001
   (i) (i) 28, 26, 24, 22, 20, 18
      (ii) Linear
      (iii) ⁻170
   (j) (i) 12, 6, 4, 3, 2.4, 2
      (ii) Non-linear
      (iii) 0.12

E Finding a formula for a linear sequence (p 313)

E1 (a) $3n + 7$ (b) $6n + 1$

E2 (a) $6n - 5$ (b) $0.5n + 3.5$

     (c) $5n - 11$ (d) $\frac{1}{4}n + 5$

E3 (a) The pupil's explanation, possibly: 'He is mixing up the formula for the nth term with the rule for going from one term to the next.'

     (b) $2n + 6$

E4 Sequence 1: constant difference = 5
Sequence 2: $6n + 5$
Sequence 3: $2n - 1$
Sequence 4: $\frac{1}{2}n + 6\frac{1}{2}$
Sequence 5: $2n - 5$
Sequence 6: 3rd term could be 10, 40, 90, 160, … (all of the form $10p^2$)

F Decreasing linear sequences (p 315)

F1 (a) $53 - 3n$

     (b) $101 - n$

F2 (a) $100 - 2n$; $^-100$

     (b) $46 - 6n$; $^-554$

     (c) $50.5 - 0.5n$; 0.5

     (d) $7.2 - 0.2n$; $^-12.8$

What progress have you made? (p 315)

1 $2n + 4$

2 (a) 10 (b) $^-2$

3 (a) $6n + 5$ (b) $9n - 7$

4 $23 - 3n$

Practice booklet

Sections A and B (p 108)

1 (a)

Pattern 4

     (b) $2n + 1$ (c) $4n - 1$

2 (a)

Pattern 4

     (b) $3n + 1$ (c) $7n - 2$

3 (a)

Pattern 4

     (b) $4n + 4$ (c) $6n + 4$

4 (a)

Pattern 4

     (b) $8n + 8$ (c) $12n + 12$

Sections C and D (p 108)

1 (a) (i) Multiply by 2
       (ii) 1536

     (b) (i) Add the previous two terms
       (ii) 191

     (c) (i) Add 2
       (ii) 112

     (d) (i) Add 1, then 2, then 3, …
       (ii) 55

     (e) (i) Subtract 2
       (ii) 7

(f) (i) Multiply by 3 and subtract 1 or
add 1, 3, 9, 27, ...
(ii) 9842

2 (a) (i) 3, 6, 9, 12, 15, 18, 21, 24
(ii) Linear
(iii) 150

(b) (i) ⁻2, ⁻1, 0, 1, 2, 3, 4, 5
(ii) Linear
(iii) 47

(c) (i) 9, 8, 7, 6, 5, 4, 3, 2
(ii) Linear
(iii) ⁻40

(d) (i) 6, 9, 14, 21, 30, 41, 54, 69
(ii) Non-linear
(iii) 2505

(e) (i) 6, 8, 10, 12, 14, 16, 18, 20
(ii) Linear
(iii) 104

(f) (i) 42, 21, 14, 10.5, 8.4, 7, 6, 5.25
(ii) Non-linear
(iii) 0.84

Section E (p 109)

1 (a) (i) $3n + 4$ (ii) 304
(b) (i) $5n + 2$ (ii) 502
(c) (i) $7n - 4$ (ii) 696
(d) (i) $5n - 5$ (ii) 495
(e) (i) $6n - 7$ (ii) 593
(f) (i) $0.75n + 4.25$
(ii) 79.25

2 (a) (i) 9, **11**, **13**, 15   (ii) $2n + 7$
(iii) 57
(b) (i) 2, **10**, **18**, 26   (ii) $8n - 6$
(iii) 194
(c) (i) ⁻1, **2**, 5, **8**   (ii) $3n - 4$
(iii) 71
(d) (i) ⁻**2**, 3, **8**, 13   (ii) $5n - 7$
(iii) 118

3 (a) 11, 15 and 19 (b) $4n + 3$

4 27th term

5 (a) £47.50   (b) £$(40 + 2.5n)$
(c) £77.50   (d) 24 weeks

Section F (p 110)

1 (a) (i) $50 - 2n$ (ii) 10
(b) (i) $21 - n$ (ii) 1
(c) (i) $105 - 5n$ (ii) 5
(d) (i) $10 - 3n$ (ii) ⁻50

2 (a) (i) 6, **4**, **2**, 0   (ii) $8 - 2n$
(iii) ⁻52
(b) (i) 15, **14**, **13**, 12   (ii) $16 - n$
(iii) ⁻14
(c) (i) ⁻2, ⁻**5**, ⁻8, ⁻**11**   (ii) $1 - 3n$
(iii) ⁻89
(d) (i) **42**, 34, **26**, 18   (ii) $50 - 8n$
(iii) ⁻190

3 (a) (i) $4n - 7$ (ii) 393
(b) (i) $9 - n$ (ii) ⁻91
(c) (i) $4n + 10$ (ii) 410
(d) (i) $15n - 5$ (ii) 1495
(e) (i) $10 - 2n$ (ii) ⁻190
(f) (i) $20 - 0.5n$ (ii) ⁻30

Ratio

p 316	**A** Stronger, darker, sweeter, happier, …	Comparisons: group discussion
p 318	**B** Ratio notation	
p 319	**C** Darker, lighter	Ordering ratios
p 320	**D** Working with ratios	Simplest form
p 322	**E** Sharing in a given ratio	
p 323	**F** Ratios in patterns	Tiling patterns
p 325	**G** Comparing ratios	
p 326	**H** Writing a ratio as a single number	

Essential

Sheet 190

Optional

Sheet 189

Practice booklet pages 111 to 115 (triangular dotty paper is needed)

Ⓐ Stronger, darker, sweeter, happier, … (p 316)

◊ The questions are meant to be exploratory, to see how pupils think about problems involving ratio and proportion. The focus is on the arguments they use.

◊ If answers are incorrect, or if pupils have no way of telling, then do not feel you have to teach the correct method here and now. There will be an opportunity to return to the questions later in the unit.

◊ There are several different strategies for each problem. Let pupils explain their methods to each other. Don't hold up one correct method as superior to others.

◊ A common error is to use differences. For example, 2 : 3 might be thought equal to 4 : 5.

◊ The answers are: **1** A; **2** A; **3** B; **4** B; **5** A; **6** the first; **7** A; **8** A; **9** A

B Ratio notation (p 318)

B3(c) If a hint is needed, point out that 3 litres of blue and 2 litres of yellow will make 5 litres of dark green.

C Darker, lighter (p 319)

> Optional:
> The recipes are printed on sheet 189 for cutting up into cards.

◊ The recipes in order from darker to lighter are (reading across):

M (4:1)   K (3:1)   B, N (4:2 = 2:1)

G (3:2)   P (4:3)   D, J, L, O (1:1 = 2:2 = 3:3 = 4:4)

C (3:4)   A (2:3)   E, F (2:4 = 1:2)   I (1:3)   H (1:4)

◊ To make a recipe between, say, 4:1 and 3:1, some pupils may say 3.5:1. This is correct but you could also ask if they can express it in whole numbers.

D Working with ratios (p 320)

E Sharing in a given ratio (p 322)

F Ratios in patterns (p 323)

> Sheet 190

◊ Pupils need to find a repeating unit from which the pattern is made. F5 is quite tricky!

G Comparing ratios (p 325)

H Writing a ratio as a single number (p 326)

There are many common instances where a ratio is given as a single number. For example the ratio of the circumference of a circle to its diameter is expressed as π rather than as $\pi:1$.

B Ratio notation (p 318)

B1 (a) 12   (b) 5

B2

Tins of blue	Tins of yellow
8	**20**
14	**35**
10	25
24	60

B3 (a) 9   (b) 10
    (c) 30 litres blue, 20 litres yellow

B4 (a) 15 litres blue, 10 litres yellow
    (b) 9 litres blue, 3 litres red
    (c) 8 litres blue, 4 litres red

D Working with ratios (p 320)

D1 (a) 15 litres   (b) 10 litres

D2 (a) 15 litres   (b) 4 litres

D3 (a) 1:2   (b) 12   (c) 20

D4 4:1

D5 4:3

D6 (a) 4:3   (b) 3:4

D7 (a) 3:1 (b) 3:2 (c) 2:3 (d) 4:7

D8 **1:5** = 4:20 = 20:100 = 2:10
    4:6 = **2:3** = 10:15
    3:1 = 9:3 = 12:4
    3:2 = 12:8

D9 10 litres blue, 20 litres yellow

D10 5 litres red, 15 litres white

*D11 (a) 15 litres
    (b) 5 litres black, 25 litres white
    (c) 4 litres black, 10 litres white
    (d) 30 litres
    (e) 15 litres black, 25 litres white

E Sharing in a given ratio (p 322)

E1 Stuart £8, Shula £4

E2 Dawn £15, Eve £5

E3 Beric £9, Betty £12

E4 (a) £16, £4   (b) £24, £36
    (c) £15, £9   (d) £25, £20
    (e) £7.50, £5   (f) £4.50, £1.50
    (g) £10, £7.50   (h) 80p, £1

E5 (a) James £750, Sarah £1250
    (b) 1 year: James £800   Sarah £1200
        2 years: James £833+ Sarah £1166+
        3 years: James £857+ Sarah £1142+
        4 years: James £875   Sarah £1125, etc
    (c) James's share goes up as grandmother lives longer.

E6 Alan £50, Bertha £100, Cyril £750

E7 (a) £3, £4.50, £6
    (b) £11, £16.50, £38.50

F Ratios in patterns (p 323)

F1 2:1

F2 (a) 2:1   (b) 3:2   (c) 1:1

F3 1:1

F4 (a) 2:1 ▦▦▢
    (b) 3:2 (▦▦▦▢▢ downwards,
       or ▦▦▢▦▢ across)
    (c) 1:3 ▦▢
          ▢▢
    (d) 4:5

F5 A possible repeating unit is shown in each case.

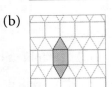

    (a)   1:1

    (b)   1:2

(c) 1:2

(d) 1:8

(e)
    (i) 1:3
    (ii) 1:2
    (iii) 2:3

Ⓖ Comparing ratios (p 325)

G1 Royal

G2 (a) $3:5 = \mathbf{15}:25$ (b) $7:8 = \mathbf{28}:32$
(c) $11:15 = \mathbf{66}:90$ (d) $4:9 = 36:\mathbf{81}$

G3 Squirrel grey   black : white = 39:15
Thundercloud grey black : white = 40:15
Thundercloud grey is darker.

G4 Jasmine's   juice : water = 55:88
Nita's   juice : water = 56:88
Nita's is stronger (just!).

G5 Initial ratio
        juice : water = 14:9 = 154:99
Final ratio
        juice : water = 17:11 = 153:99
The final drink was weaker (just!).

Challenge
Red : green = 3:5 = 12:20
Green : blue = 4:7 = 20:35
So red : blue = 12:35

Ⓗ Writing a ratio as a number (p 326)

H1 (a) 2   (b) 1.6 (c) 1.5 (d) 0.75

H2 The window is square.

H3 Hamster 0.2   Cat 0.1   Elephant 0.004

H4 A 0.44…   B 0.4…   A is steeper.

What progress have you made? (p 327)

1 (a) 20 litres   (b) 6 litres

2 (a) 2:5   (b) 2:3

3 Gavin £16, Susan £20

4 Peter £12.50, Paul £37.50, Mary £50

5 A black : white = 25:40
    B black : white = 24:40
    A is darker.

6 0.16

Practice booklet

Section B (p 111)

1 (a) (i) 15 g   (ii) 36 g
    (b) (i) 10 g   (ii) 14 g

2

Pure gold (g)	Other metals (g)
6	**10**
12	**20**
15	**25**
9	15
18	30
45	75

3 (a) 6 g silver, 3 g copper
    (b) 10 g copper, 330 g gold
    (c) 495 g gold, 15 g copper

Section D (p 112)

1 (a) 4:3   (b) 3:4

2 (a) 2:3   (b) 3:1   (c) 3:5
    (d) 5:2   (e) 7:6

3 (a) 2:3   **6:9**   10:**15**
    (b) 4:1   **12:3**   8:**2**
    (c) 6:2   **3:1**   18:**6**
    (d) 4:10 2:**5**   **6**:15
    (e) 30:40 **6:8**   3:**4**
    (f) 7:3   21:**9**   **28**:12

4 20:30 and 6:9
21:6 and 7:2
9:3 and 12:4
3:5 and 12:20
12:9 and 8:6
8:20 and 2:5

5 (a) 9   (b) 4
(c) 15 loam and 5 sand

6 (a) 4   (b) 12
(c) 15 blue and 10 red

Section E (p 113)

1 (a) £4 and £6   (b) £10 and £15
(c) £32 and £48   (d) £17 and £25.50

2 (a) £10   (b) £12
(c) £18.75 and £11.25

3 (a) 20 litres blue, 16 litres green
(b) 7.5 litres blue, 6 litres green
(c) 9.5 litres blue, 7.6 litres green

4 (a) £20 £40 £60   (b) £48 £24 £48
(c) £24 £36 £60   (d) £20 £30 £70

5 (a) £0.50 £1   £2
(b) £0.70 £1.05 £1.75
(c) £0.75 £1.25 £1.50
(d) £1   £1.20   £1.30

Section F (p 114)

1 (a) 3:4   (b) 2:3
(c) 3:1   (d) 3:5

2 One way to do this is to take a row of 10 triangles and colour 6 of them and leave 4 blank. If this block is repeated, the pattern will have the desired ratio of grey to white.

Section G (p 115)

1 A sheets : envelopes = 40:15 = 120:45
B sheets : envelopes = 25:9 = 125:45
B has the higher ratio.

2 Rough puff, Flan, Choux, Flaky, Puff

Section H (p 115)

1 (a) 8 cm by 13 cm
13 cm by 21 cm
21 cm by 34 cm

(b) A 2
B 1.5
C 1.66…
D 1.6
The next three rectangles give
1.625   1.615…   1.619…

(c) The answers alternate smaller and bigger. They get closer to 1.618 033… , the golden ratio.

40 Using a spreadsheet

A number of activities relating to different topic areas are gathered together here. If your class has easy access to computers, then you could dip in whenever appropriate. If, however, you have access only at specific times, you could use the unit as a programme of work for your computer sessions. Alternatively, you could ask for the activities to be used in IT lessons which focus on the use of a spreadsheet.

Essential

Spreadsheet, or similar graphic calculator facility

Spot the formula (p 328)

◊ Some pupils may invent formulas that are impossibly difficult to spot.

◊ Opportunities to discuss equivalent formulas may arise, for example = (A1 + 3)*2 and = 2*A1 + 6.

◊ It is interesting to discuss strategies. For example, putting 100 or 1000 into a formula can often tell you a lot.

Making a sequence (p 329)

◊ You may need to introduce or revise the process of filling down a formula (or 'drag and drop').

◊ Pupils may find a formula for going from one term to the next. Although this is valid, get them to focus on finding a formula that works **across**, calculating the terms of the sequence from the numbers 1, 2, 3, …

Big, bigger, biggest (p 329)

The idea of using decimals may not occur at first.
Here are some solutions (others are possible in some cases):

1 (a) 2, 2, 3, 7   (b) 2, 3, 3, 6   (c) 3.5, 3.5, 3.5, 3.5 give 150.0625
2 (a) 3, 3, 4, 5 or 2.5, 4, 4, 4.5   (b) 2, 2.5, 5, 5.5
   (c) 3.75, 3.75, 3.75, 3.75 give 197.753 906 25

Ways and means (p 330)

Pupils will soon see that the sequence tends to a limit. The challenge is to find how the limit is related to the starting numbers. For two starting numbers a, b the limit is $\frac{1}{2}(a + 2b)$, for three starting numbers a, b, c it is $\frac{1}{6}(a + 2b + 3c)$, and for four, a, b, c, d, it is $\frac{1}{10}(a + 2b + 3c + 4d)$.

Parcel volume (p 330)

The main focus here is on working systematically. As with many problems of this type it helps to restrict the number of variables. If the spreadsheet is set up in the most obvious way (as shown in the pupil's book), it may be difficult to keep track of the restrictions on the variables.

Ask pupils how they could build in the connection between length and girth (which must add up to 3 m for maximum volume). One way is to use a formula equivalent to '3 – girth' in the 'length' column. As girth is already defined as 2(height + width), this reduces the variables to two: height and width.

The maximum volume of 0.25 m^3 occurs when the width and height are both 0.5 m and the length is 1 m.

Furry festivals (p 331)

Pupils may themselves suggest reducing the number of variables by setting the number of pens equal to 60 minus the total number of T-shirts and badges.

The solution is:

    5 T-shirts, 35 badges and 20 pens

 or 6 T-shirts, 14 badges and 40 pens

Breakfast time (p 331)

The essential step is to set up a formula for working out a given percentage of a quantity. Pupils may know that, for example, 8 in the percentage column has to be converted to 0.08 but may not immediately realise that the operation needed to do this is ÷ 100.

Mean, median, range (p 332)

In this activity pupils investigate the effect on the mean, median and range of changes in individual data items or transformations of all the data items.

Solution to question 4:

(a) You could increase either 2, 4, 6, 13, 15 or 17 by 1.

(b) Change 13 to 24.

(c) Change 11 to 2, or 13 to 4, or 15 to 6.

(d) Change 10 to 22.

Functions and graphs 7S/51

Essential	Optional
2 mm graph paper	OHP
Practice booklet pages 116 and 117 (graph paper needed)	

Ⓐ **From table to graph** (p 333)

> 2 mm graph paper

◊ The main points to be covered in the introduction are:
  - We can make a table from a rule given in words.
  - The values in the table can be plotted as points.
  - When the variables are continuous, a line can be drawn through the points to show the relationship between the variables.

Ⓑ **Functions** (p 335)

Ⓒ **Spot the function** (p 336)

> Optional: OHP

This is similar to the activity in unit 19 'Spot the rule'. You will need a large grid on the board or OHP with x- and y-axes.

◊ You can start the activity by telling the class that you are thinking of a rule linking x and y, for example $y = x + 2$. Ask a member of the class to give you a value of x; work out y and plot the corresponding point on the grid. Continue until someone can tell you your rule.

Then ask a pupil to take over with a rule of their own.

◊ If pupils are adventurous and use rules involving, for example, squaring, then you can ask what kinds of rule give points which lie in a straight line. (If nobody thinks of squaring, then do so yourself to bring out the distinction between linear and non-linear functions and to explain why the word 'linear' is used for rules of this kind.)

◊ Again, if nobody else does so, use a rule like $x + y = 10$, and bring out the fact that this can also be written $y = 10 - x$ or as $y = {}^{-}x + 10$.

◊ Ask pupils if they see any similarities and differences between earlier work on sequences and this work. They should be able to appreciate that
- in the case of a sequence, n is restricted to being a whole number (discrete), whereas x is continuous
- the values of y for $x = 1, 2, 3, 4, \dots$ form a sequence (which helps when pupils are trying to find the equation of a linear graph)

C7 You should point out that in real life relationships are rarely precisely linear (except in cases where linearity is built in, as for example with currency conversion graphs).

Ⓐ **From table to graph** (p 333)

A1 (a)

Gas in tank (kg)	Hours away from base
3	5
4	7
5	9
6	11
7	13
8	15
9	17

(b) The pupil's graph of the table, points joined with a line

(c) 8 hours

(d) $5\frac{1}{2}$ kg

(e) 199 hours

A2 (a)

t	0	1	2	3	4	5	6
h	20	50	80	110	140	170	200

(b) $h = 30t + 20$

(c) The pupil's graph of points from the table

(d) $t = 2.7$ (roughly)

A3 (a)

w	0	2	4	6	8	10
d	100	90	80	70	60	50

(b) The pupil's graph of points from the table, labelled '$d = 100 - 5w$'

(c) $w = 4.6$

(d) 7 tonnes

Ⓑ **Functions** (p 335)

B1 (a)

x	⁻6	⁻4	⁻2	0	2	4	6
y	⁻5	⁻3	⁻1	1	3	5	7

(b) The pupil's graph of $y = x + 1$, labelled

B2 (a)

x	⁻1	0	1	2	3	4
y	⁻5	⁻3	⁻1	1	3	5

(b) The pupil's graph of $y = 2x - 3$, labelled

B3 (a)

x	⁻4	⁻2	0	2	4	6
y	6	4	2	0	⁻2	⁻4

(b) The pupil's graph of $y = 2 - x$, labelled

B4 (a) The pupil's table for $y = 5 - 2x$

(b) The pupil's labelled graph of $y = 5 - 2x$

Ⓒ **Spot the function** (p 336)

C1

x	0	1	2	3	4
y	2	5	8	11	14

$y = 3x + 2$

C2

x	0	1	2	3	4
y	10	9	8	7	6

Paula

C3 $y = x + 4$

C4 Prakesh

C5 $y = x - 10$

C6 (a) $y = x + 2$ (b) $y = x - 4$
     (c) $y = 8$
     (d) $y = 6 - x$ or $x + y = 6$

C7 $L = 5W + 20$

C8 $d = 350 - 50t$

*C9 (a) $y = 3x - 5$ (b) $y = 10 - 2x$
     (c) $y = \frac{1}{2}x + 2\frac{1}{2}$

*C10 (a)

C	⁻15	⁻10	⁻5	0	5	10	15	20
F	5	14	23	32	41	50	59	68

(b) and (c)

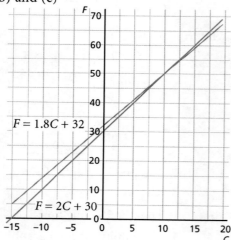

(d) $C = 10$

(e) ⁻15 to 35 (35 is as far above 10 as ⁻15 is below)

What progress have you made? (p 339)

1 The pupil's graph of $y = 2x + 3$

2 (a) $y = 2x + 2$
    (b) $y = 7 - x$ or $x + y = 7$

Practice booklet

Section A (p 116)

1 (a)

t	0	1	2	3	4	5
d	80	100	120	140	160	180

  (b) $d = 80 + 20t$
  (c) The pupil's graph from the table above
  (d) $t = 2.4$

2 (a)

t	0	1	2	3	4	5
d	30	26	22	18	14	10

  (b) The pupil's graph from the table above
  (c) 3.5 minutes

Sections B and C (p 117)

1 (a)

x	⁻4	⁻3	⁻2	⁻1	0	1	2	3	4
y	⁻12	⁻10	⁻8	⁻6	⁻4	⁻2	0	2	4

  (b) The pupil's labelled graph of $y = 2x - 4$

2 (a)

x	⁻4	⁻3	⁻2	⁻1	0	1	2	3	4
y	17	15	13	11	9	7	5	3	1

  (b) The pupil's labelled graph of $y = 9 - 2x$

3 (a) $y = x + 7$
  (b) $x + y = 2$ or $y = 2 - x$
  (c) $y = 3x - 3$

4 $h = 35 - 5w$

Review 5 (p 340)

1 (a) 1.215 44   (b) 1.555
  (c) 7.046 718 75

2 (a) 87.75 cm^3 (b) 144 cm^2

3 (a) (i) $6n - 3$ (ii) $50 - 4n$
  (b) (i) 117   (ii) $^-30$

4 (a) $10 + 6a$ (b) $24b^2$ (c) $^-2c - 5$
  (d) Cannot be simplified
  (e) 3   (f) $14f^2$

5 Jane: 65.2% (to the nearest 0.1%)
 Sinead: 63.3% (to the nearest 0.1%)
 Jane did better.

6 8

7 (a)

t	0	1	2	3	4	5	6
d	25	23	21	19	17	15	13

  (b) The pupil's straight line graph
  (c) $d = 25 - 2t$ (d) $12\frac{1}{2}$ days

8 (a) 2:3 (b) 2:3 (c) 2:3
  (d) 81, 121.5

9 (a) $n = 9$ (b) $n = ^-7$ (c) $n = 140$

10 (a) $y = x - 4$ (b) $y = 2x + 4$
  (c) $x + y = 4$

11 (a) 5:4   (b) 5:11

12 $\frac{1}{4}$ 0.3 $\frac{8}{25}$ $\frac{17}{50}$ $\frac{7}{20}$ 0.37 $\frac{39}{100}$ $\frac{2}{5}$

13 (a) $x = 32°$ (b) $y = 124°$ (c) $z = 95°$
  with pupil's reasons

14 (a) The totals are equal.
  (b) The pupil's addition crosses
  (c) Sum of red squares
    $= a + (b + c) = a + b + c$
    Sum of blue squares
    $= b + (a + c) = a + b + c$
  (d) The products of the 'red' and 'blue'
    squares are equal.
    Both are equal to $a \times b \times c$

15 $1 \times 1 \times 60$
  $1 \times 2 \times 30$
  $1 \times 3 \times 20$
  $1 \times 4 \times 15$
  $1 \times 5 \times 12$
  $1 \times 6 \times 10$
  $2 \times 2 \times 15$
  $2 \times 3 \times 10$
  $2 \times 5 \times 6$
  $3 \times 4 \times 5$

16 Area of garden $= 16^2$ m^2 $= 256$ m^2
  Area of lawn $= 256 \div 2 = 128$ m^2
  Length of each side of lawn
  $= \sqrt{128} = 11.31...$ m
  Width of paths
  $= \frac{1}{2}(16 - 11.31...) = 2.34$ m (to 2 d.p.)

Mixed questions 5 (Practice booklet p 118)

1 (a) 528 m^2 (b) 63 360 m^3
  (c) 768 lorry loads

2 (a) 14 cm^2, 30 cm^2 (b) $4n + 2$
  (c) 37 cubes   (d) $16n + 6$

3 (a) Ann £7.50, Brian £10, Charlie £12.50
  (b) Ann £8, Brian £10, Charlie £12

4 (a) $3n - 1$ (b) $4n - 11$
  (c) $28 - 2n$ (d) $14 - 9n$

5 (a) $y = x + 5$ (b) $y = x - 4$
  (c) $y = 3 - x$ (d) $y = 2$
  (e) $y = 2x + 3$

6 (a) (i) 105   (ii) $^-95$
  (b) (i) 96   (ii) $^-104$
  (c) (i) $^-97$ (ii) 103
  (d) (i) 2   (ii) 2
  (e) (i) 203   (ii) $^-197$

7 (a) 2.76   (b) 3.85

8 25%

9 (a) 1 person   (b) 2 people

(c) 2.48 people

10 (a) $^-32$ (b) $^-32 + 8(n - 1)$

11 (a) RL = red left   RR = red right

BL = blue left   BR = blue right

GL = green left   GR = green right

$\otimes$ = L and R choices

2nd glove: GR, GL, BR, BL, RR, RL

1st glove: RL RR BL BR GL GR

(b) $\frac{9}{15} = \frac{3}{5}$ (c) $\frac{3}{15} = \frac{1}{5}$

*12 9:31